Algebraic Structures and Operator Calculus

Mathematics and Its Applications

Managing Editor:

M. HAZEWINKEL

Centre for Mathematics and Computer Science, Amsterdam, The Netherlands

Volume 241

Algebraic Structures and Operator Calculus

Volume I:
Representations and Probability Theory

by

Philip Feinsilver

Department of Mathematics,
Southern Illinois University,
Carbondale, Illinois, U.S.A.

and

René Schott

CRIN,
Université de Nancy 1,
Vandoeuvre-les-Nancy, France

SPRINGER SCIENCE+BUSINESS MEDIA, B.V.

Library of Congress Cataloging-in-Publication Data

Feinsilver, Philip J., 1948-
   Algebraic structures and operator calculus / by Philip Feinsilver
and René Schott.
      p.   cm. -- (Mathematics and its applications ; v. 241)
   Includes bibliographical references and index.
   Contents: v. 1. Representations and probability theory
   ISBN 978-94-010-4720-3   ISBN 978-94-011-1648-0 (eBook)
   DOI 10.1007/978-94-011-1648-0
   1. Operational calculus. 2. Probabilities. 3. Representations of
groups.   I. Schott, René. II. Title. III. Series: Mathematics and
its applications (Kluwer Academic Publishers) ; v. 241.
QA432.F45   1993
515'.72--dc20   92-44824

ISBN 978-94-010-4720-3

Printed on acid-free paper

To Micheline and Sylvia

Table of Contents

Chapter 5 Bernoulli Processes

Chapter 6 Bernoulli Systems

Chapter 7 Matrix Elements

Preface

This series presents some tools of applied mathematics in the areas of probability theory, operator calculus, representation theory, and special functions used currently, and we expect more and more in the future, for solving problems in mathematics, physics, and, now, computer science. Much of the material is scattered throughout available literature, however, we have nowhere found in accessible form all of this material collected. The presentation of the material is original with the authors. The presentation of probability theory in connection with group representations is new, this appears in Volume I. Then the applications to computer science in Volume II are original as well. The approach found in Volume III, which deals in large part with infinite-dimensional representations of Lie algebras/Lie groups, is new as well, being inspired by the desire to find a recursive method for calculating group representations. One idea behind this is the possibility of symbolic computation of the matrix elements.

In this volume, *Representations and Probability Theory*, we present an introduction to Lie algebras and Lie groups emphasizing the connections with operator calculus, which we interpret through representations, principally, the action of the Lie algebras on spaces of polynomials. The main features are the connection with probability theory via moment systems and the connection with the classical elementary distributions via representation theory. The various systems of polynomials that arise are one of the most interesting aspects of this study.

As a text the work is intended for senior and first year graduate level students. It is suitable for self-study for researchers as well. Each chapter has exercises ranging in difficulty from fairly easy to quite advanced. The reader should feel free to skip an exercise, but we emphasize the importance of working through details so that a real facility for working with the basic noncommutative algebras and hypergeometric functions is developed. It will be noted that we do not have many exercises in basic probability theory as these are available in standard texts.

We would like to acknowledge our colleagues at the University of Nancy — the CRIN-INRIA Lorraine and the Department of Mathematics — whose support and encouragement is truly appreciated. Especially, the first author is grateful for support during several visits, including a sabbatical visit, that enabled us to carry out this work. We would like as well to acknowledge the support of the Department of Mathematics, Southern Illinois University at Carbondale. The second author especially would like to express his appreciation for the hospitality and pleasant working environment during several visits made to the department. In addition to our teachers and colleagues, we mention Profs. R.A. Askey and G.-C. Rota whose encouragement over many years is particularly appreciated.

INTRODUCTION

I. General remarks

We will present an original approach to probability theory through operator algebras — representations of Lie algebras. Along the way we will meet many special functions, particularly orthogonal polynomials. Principal roles will be played by the Heisenberg-Weyl algebra and the sl(2) algebra. The oscillator algebra will appear as well. These are introduced in *Chapter 1* where basic techniques for working with noncommutative variables are presented. Formulas for the groups corresponding to the various algebras are discussed there. *Chapter 2* is a brief foray into the subject of hypergeometric functions. Here some illustrations of our algebraic approach are to be found. *Chapter 3* is a presentation of probability theory from the operator algebra point of view — often referred to as 'quantum probability'. The results are stated in the finite-dimensional case, but they are formulated (and later used) in the infinite-dimensional case, so that the presentation is applicable for the general theory. Fock spaces are introduced and connections with analytic structure, e.g. reproducing kernels, are given. Clebsch-Gordan coefficients are calculated, with connections to orthogonal polynomials.

Chapter 4 presents the theory of moment systems. These are polynomial solutions to evolution equations — polynomials arising from the powers x^n by convolution with a semigroup of probability measures, and we include extensions to general types of Appell polynomials. Systems related to the algebra sl(2), radial moment systems, are presented as well. The chapter closes with the theory of change-of-variables for the HW algebra. *Chapter 5* presents Bernoulli processes. These are processes for which there are corresponding moment systems (in HW variables) that yield systems of orthogonal polynomials as well. Their role as bases for representations of the sl(2), oscillator, and HW algebras is a primary feature. This shows that these fundamental stochastic processes have a natural foundation in Lie groups. *Chapter 6* presents the general theory of Bernoulli systems. This approach is quite general and indicates the formulation used for processes in higher dimensions, i.e., vector-valued processes. The chapter closes with the L^2 reproducing kernels for the processes of Chapter 5. *Chapter 7* presents calculations of the matrix elements for the groups that have played a role throughout. Coherent state representations are discussed; their significance from the mathematical point of view is explained. The matrix elements yield addition formulas for the special functions that arise in the representations. The confluence of the major functions that are primary in the theory of Bernoulli systems in the setting of addition formulas for the matrix elements concludes the work.

We have paced the material for the 'working reader' — who wants to be able really to use it. That is, the reader will find most arguments in proofs not to include all detailed calculations. These should be considered as exercises. In many cases indications are given explicitly in the list of exercises to fill in details of proofs, etc. There are additional exercises and examples involving various approaches to the material, illustrations from other points of view, etc., etc. Our intention is to provide the guide to 'calculus' for those for whom Lie groups and representations, noncommutative mathematics in general, will be in daily employ.

II. Some notations

1. For hypergeometric functions, we use the standard notations:

$$(a)_b = \frac{\Gamma(a+b)}{\Gamma(a)}$$

(see §III below)

It is also useful to use the notation for factorial powers, for integer $n > 0$, with $a^{(0)} = 1$

$$a^{(n)} = a(a-1)\cdots(a-n+1) = (-1)^n(-a)_n$$

with the usual binomial coefficient

$$\binom{a}{k} = \frac{a^{(k)}}{k!}$$

A useful symbol that arises naturally in the commutation formulas is a slight extension of the binomial coefficient:

$$\binom{m,n}{k} = \binom{m}{k}\binom{n}{k}k! = \frac{m!\,n!}{(m-k)!\,(n-k)!\,k!}$$

2. The symbol i will be reserved to mean $\sqrt{-1}$. The letters j,k,l,m,n will be understood always to denote integers; in fact, nonnegative integers unless indicated otherwise, e.g. by a summation from $-\infty$ to ∞.

3. Generally in statements of propositions summations will be written out using Latin indices. It is useful to employ the following *summation convention* (especially for calculations in deriving formulas), namely: *repeated Greek indices are summed over, regardless of position*. E.g., $\sum x_j y_j = x_\lambda y_\lambda$, $e^x = x^\alpha/\alpha!$, and so on. Latin indices are summed over *only* when there is a corresponding summation sign.

4. We conventionally use Kronecker's delta δ_{jk} such that

$$\delta_{jk} = \begin{cases} 1, & \text{if } j = k \\ 0, & \text{otherwise} \end{cases}$$

and the Levi-Civita totally antisymmetric symbol ε_{jkl} for indices $1 \leq j, k, l \leq 3$, where:

$$\varepsilon_{jkl} = \begin{cases} 1, & \text{if } j = 1, \, k = 2, \, l = 3 \\ 1, & \text{if the indices } jkl \text{ are an even permutation of 123} \\ -1, & \text{if the indices } jkl \text{ are an odd permutation of 123} \end{cases}$$

That is, ε_{jkl} changes sign every time any two indices are interchanged, starting with $\varepsilon_{123} = 1$.

5. For complex numbers, we use the notations for real part and imaginary part of $z \in \mathbf{C}$: Re z, Im z. For example,

$$z = x + iy, \qquad (x, y) \in \mathbf{R}^2 : \qquad x = \mathrm{Re}\, z, \, y = \mathrm{Im}\, z$$

6. All italicized terms are to be found in the index. This includes terms defined in formal Definitions as well as those defined in the text.

III.   Gamma and beta functions

The *gamma function* may be defined by the integral

$$\Gamma(\alpha) = \int_0^\infty x^{\alpha-1} e^{-x} \, dx, \qquad \mathrm{Re}\, \alpha > 0$$

We recall the following without proof (see, e.g., Rainville[40]):

3.1 Proposition.   *Properties of the gamma function:*

1. *The functional equation:*
$$\alpha \, \Gamma(\alpha) = \Gamma(\alpha + 1)$$

2. *Relation with factorials:*
$$\Gamma(n) = (n - 1)!$$
   for $n \geq 1$.

3. *For half-integer factorials, we have the special value:*
$$\Gamma(\tfrac{1}{2}) = \sqrt{\pi}$$

Remark. First, recall the notation $(\alpha)_n$:

$$(\alpha)_n = \frac{\Gamma(\alpha + n)}{\Gamma(\alpha)} = \alpha(\alpha + 1) \cdots (\alpha + n - 1)$$

with the last equality following from the functional equation for integer $n > 0$.
Next, we list some elementary relations that are especially useful when dealing
with binomial coefficients and hypergeometric functions.

$$\left(\frac{1}{2}\right)_n = \frac{(2n)!}{4^n n!}, \qquad (A)_{2n} = 4^n \left(\frac{A}{2}\right)_n \left(\frac{1+A}{2}\right)_n \qquad (3.1)$$

$$(A)_{n-k} = \frac{(-1)^k (A)_n}{(1-A-n)_k}, \qquad (A+k)_{n-k} = \frac{(A)_n}{(A)_k} \qquad (3.2)$$

$$(-1)^k (A)_k = (1-A-k)_k = k! \binom{-A}{k} = (-1)^k k! \binom{A+k-1}{k} \qquad (3.3)$$

The *beta function* is given by the integral

$$B(\alpha, \beta) = \int_0^1 t^{\alpha-1} (1-t)^{\beta-1} dt$$

for $\operatorname{Re} \alpha, \operatorname{Re} \beta > 0$. We summarize the principal properties we need (Rainville[40]):

3.2 Proposition.   *Properties of the beta function:*

1. *Evaluation in terms of the gamma function:*

$$B(\alpha, \beta) = \frac{\Gamma(\alpha)\Gamma(\beta)}{\Gamma(\alpha+\beta)}$$

2. *Alternate expression as an integral:*

$$B(\alpha, \beta - \alpha) = \int_0^\infty \frac{t^{\alpha-1}}{(1+t)^\beta} dt$$

And we have the following integral whose evaluation involves the beta function:

3.3 Proposition.   *For $\alpha, \beta > 0$:*

$$\frac{2}{\pi} \int_0^{\pi/2} \cos^{2\alpha} \theta \sin^{2\beta} \theta \, d\theta = \frac{\left(\frac{1}{2}\right)_\alpha \left(\frac{1}{2}\right)_\beta}{\Gamma(\alpha+\beta+1)}$$

and for integers $\alpha, \beta \geq 0$, the integral has the value:

$$\frac{1}{4^{\alpha+\beta}} \binom{2\alpha}{\alpha} \binom{2\beta}{\beta} \bigg/ \binom{\alpha+\beta}{\alpha}$$

Proof: The substitution $t = \sin^2 \theta$ leads to the beta function $B(\alpha + \frac{1}{2}, \beta + \frac{1}{2})$.
Evaluating in terms of the gamma function leads to the stated results.   ∎

IV.   Numbering

Each chapter has independently numbered statements and equations. A statement is any Proposition, Definition, Theorem, Lemma, or Corollary. These are numbered consecutively within each (sub)section. If a reference is to another chapter, it is explicitly noted. Within chapters, there are headings with roman numerals, main sections, then sections and subsections with arabic numerals, indicating the heading and then appropriate (sub)section: e.g., section 4.3 is the third section under heading IV. Statements and equations are numbered similarly. To illustrate:

3.2 Proposition.   means the second statement of main section (heading) III

4.1.4 Proposition   would be the fourth statement, heading IV, section 4.1

5.2.4.1 Theorem   would be the first statement, heading V, section 5.2, subsection 5.2.4

This makes it very useful for browsing within chapters.

V.   References

A full list of references is found at the back of the book. Within each chapter, some references are given that relate particularly to the material under discussion. General references to a given chapter are given just before the exercises for that chapter.

VI.   Polynomials: notations and formulas

We conclude this introduction with a reference page listing the classes of polynomials which we will encounter and the notations employed. The formulas given here are expressed in terms of binomial coefficients. For the 'polynomials of Bernoulli type' the formulation in terms of hypergeometric functions is given in Ch. 5, with general formulas given in Ch. 5, §1.5.

REFERENCE LIST FOR POLYNOMIALS

The Gegenbauer polynomials make their first appearance in Ch. 3, §2.4. The Hahn polynomials appear in Ch. 3, §3.2. Tchebychev polynomials appear in this volume in the examples (§7.2 below) and play a role in volume 2. Hankel moment polynomials are defined in Ch. 4, §3.2.2. (They appear implicitly in an important role in Ch. 3, §2.4.2.) They are related to Jacobi and Gegenbauer polynomials. Jacobi polynomials appear as vacuum states in Ch. 3, §3.2. The rest are of 'Bernoulli type' and are discussed in detail in Ch. 5 (although the Krawtchouk polynomials are to be found in Ch. 3, §3.1 as well). Their specific normalizations correspond to the general theory of Ch. 5 and vary from author to author.

Meixner-Pollaczek:

$$B_n(x,t) = n! \, (i \sec \vartheta)^n e^{-in\vartheta} \sum_k \binom{-X}{k} \binom{X-t}{n-k} (-1)^k e^{2ki\vartheta}$$

with $X = (t - ix)/2$.

Gegenbauer (ultraspherical):

$$C_n^\nu(x) = \sum_k \frac{(-1)^{n+k}(\nu)_k}{(2k-n)!\,(n-k)!} \, (2x)^{2k-n}$$

Hermite:

$$H_n(x,t) = \sum_k \binom{n}{2k} \frac{(2k)!}{2^k k!} (-1)^k x^{n-2k} t^k$$

Hahn:

$$Ha_n(x) = \sum_k \binom{N-x}{n-k} \binom{x}{k} \binom{c_1+n-1}{k} \binom{c_2+n-1}{n-k} k!(n-k)!(-1)^k$$

Bernoulli, general form:

$$J_n(x,t) = n! \sum_k \binom{-X}{k} \binom{X-t}{n-k} (\alpha+\delta)^{n-k}(\alpha-\delta)^k$$

with $X = (x + \delta t)/2\delta$.

Krawtchouk:
$$K_n(x, N) = n! \sum_k \binom{x}{k} \binom{N-x}{n-k} (-1)^k p^k q^{n-k}$$

Laguerre:
$$L_n(x, t) = \sum_k \binom{n}{k} \binom{t+n-1}{n-k} (n-k)! (-1)^{n-k} x^k$$

Meixner:
$$M_n(x, N) = n! \, p^{-n} \sum_k \binom{-x-N/2}{k} \binom{x-N/2}{n-k} q^k$$

Poisson-Charlier:
$$P_n(x, t) = \sum_k \binom{n}{k} (-1)^{n-k} x^{(k)} t^{n-k}$$

Jacobi:
$$P_n^{(\alpha,\beta)}(x, t) = \frac{1}{2^n n!} \sum_k \binom{n}{k} \frac{(\alpha+1)_n (\beta+1)_n}{(\alpha+1)_{n-k} (\beta+1)_k} (x-1)^{n-k} (x+1)^k$$

Tchebychev:
$$T_n(x) = \tfrac{1}{2} \sum_k \binom{n-k}{k} \frac{n}{n-k} (-1)^k (2x)^{n-2k}$$

Tchebychev polynomials of the second kind:
$$U_n(x) = \sum_k \binom{n-k}{k} (-1)^k (2x)^{n-2k}$$

Hankel moment:
$$\Phi_n(x) = \sum_k \binom{n}{k} \frac{(c)_n}{(c)_{n-k} (c)_k} x^k$$

VII.   Exercises and examples

7.1 EXERCISES

1. Derive equations (3.1)–(3.3).

2. Prove statement #2 of Proposition 3.2.

3. Fill in the steps of the proof of Proposition 3.3.

7.2 TCHEBYCHEV POLYNOMIALS

The *Tchebychev polynomials* are defined by the relation

$$T_n(\cos\theta) = \cos n\theta$$

With $x = \cos\theta$, we have the Tchebychev polynomials defined on the interval $[-1,1]$.

1. The T_n satisfy the recurrence formula

$$2xT_n = T_{n+1} + T_{n-1}$$

   with $T_0 = 1$, $T_1 = x$. Use the recurrence formula to find T_n for $2 \le n \le 5$.

2. Derive the formula given in the Reference List.

3. Let $u_n(x,y) = r^n T_n(x/r)$, with $r = \sqrt{x^2 + y^2}$, on $\mathbf{R}^2$. Show that u_n are *harmonic functions*. I.e., they satisfy Laplace's equation

$$\frac{\partial^2 u}{\partial x^2} + \frac{\partial^2 u}{\partial y^2} = 0$$

   Verify explicitly for T_n, $n = 1, 2, 3$, using your previous results.

The *Tchebychev polynomials of the second kind* $U_n(x)$ are defined by the relation

$$U_n(\cos\theta) = \frac{\sin(n+1)\theta}{\sin\theta}$$

4. The U_n have generating function

$$\frac{1}{1 - 2xv + v^2} = \sum_{n=0}^{\infty} v^n U_n(x)$$

5. Use the generating function (e.g.) to find the formula given in the Reference List. Use that formula to calculate $U_n(x)$ for $1 \le n \le 5$.

6. Show that the polynomials $v_n(x,y) = yr^n U_n(x/r)$ are harmonic on $\mathbf{R}^2$.

7. Find a three-term recurrence for $U_n(x)$, with $U_0 = 1$, $U_1 = 2x$.

Chapter 1

INTRODUCTORY NONCOMMUTATIVE ALGEBRA

I. Representations

Consider an *algebra* e.g. $n \times n$ matrices of real numbers. First, it is a *vector space* having operations of addition and multiplication by scalars. Second, it is a *ring* having an associative multiplication compatible with the addition operation, i.e., with distributive laws. We will be interested in Lie algebras. A *Lie algebra* is a vector space with a non-associative multiplication — the Lie bracket, denoted $[A, B]$ — satisfying:

$$[A, B] = -[B, A]$$
$$[A, [B, C]] + [B, [C, A]] + [C, [A, B]] = 0$$

i.e., antisymmetry and the *Jacobi identity* .

Here we will consider *representations* . Namely, where the algebra acts as linear operators on a vector space. For a Lie algebra, the abstract bracket becomes the commutator

$$[A, B] = AB - BA$$

which will be our usage throughout.

1.1 LEFT AND RIGHT MULTIPLICATION OPERATORS

First, consider $n \times n$ matrices of real or complex numbers, denoted here by $\mathcal{M}$. Let *Lin($\mathcal{M}$)* denote the linear transformations of $\mathcal{M}$. We can define mappings from $\mathcal{M}$ into Lin($\mathcal{M}$) by

$$r_A X = XA$$
$$l_A X = AX$$

Notice that l gives a representation, i.e., it is a homomorphism, of $\mathcal{M}$: $l_{AB} = l_A l_B$, the order of multiplication is preserved; while r is an anti-homomorphism: $r_{AB} = r_B r_A$, the order of multiplication is reversed. The representation l is called the *left regular representation* of $\mathcal{M}$. What we want to consider is the algebra generated by these operators $\{ l_A, r_A : A \in \mathcal{M} \}$.

1.1.1 Proposition. *For all $A, B \in \mathcal{M}$, l_A and r_B commute. I.e., $[l_A, r_B] = 0$.*

Notice that this is exactly the associativity of multiplication: $(AX)B = A(XB)$.

A feature that is evident from this example is that the dimensions of representation spaces are not obvious. E.g., we started out with $n \times n$ matrices, which are naturally considered as acting on n-dimensional space, and are looking at a representation on an n^2 dimensional space, $\mathcal{M}$ itself. A common terminology is to call the dimension of the representation space the *degree* of the representation.

Remark. For this work, A, B, X, etc. will generally denote elements from an associative algebra.

1.2 COMMUTATION RELATIONS AND BASIC FORMULAS

The commutativity of the left and right multiplication operators leads readily to some basic commutation formulas. First, note that the usual geometric series and binomial identities hold:

1.2.1 Proposition. *For any A and B,*

$$l_A^n - r_B^n = (l_A - r_B)(l_A^{n-1} + l_A^{n-2} r_B + \cdots + r_B^{n-1})$$
$$(l_A + r_B)^n = \sum_{k=0}^{n} \binom{n}{k} l_A^{n-k} r_B^k$$

Applying the first line of this Proposition yields:

1.2.2 Proposition.

$$A^n X - X B^n = \sum_{j=0}^{n-1} A^{n-1-j}(AX - XB)B^j$$

$$X A^n - B^n X = \sum_{j=0}^{n-1} B^{n-1-j}(XA - BX)A^j$$

Similarly, the binomial identity yields:

1.2.3 Proposition. *Define the operator Θ: $\Theta(X) = AX + XB$. Then*

$$\Theta^n(X) = \sum_{k=0}^{n} \binom{n}{k} (A^{n-k} X B^k)$$

In particular, when $B = -A$, this operator is denoted by ad A, i.e., $(\text{ad }A)(X) = AX - XA = [A, X]$.

1.3 EXPONENTIALS AND GROUP ELEMENTS

Corresponding to elements of a Lie algebra are *group elements* via exponentiation. Given A, a matrix, the series

$$e^{tA} = \sum_{n=0}^{\infty} \frac{t^n A^n}{n!}$$

is well defined. For $-\infty < t < \infty$, $U(t) = e^{tA}$, gives a group of operators. They satisfy

$$U(t+s) = U(t)U(s), \qquad U(0) = I \quad \text{identity operator}$$

$$\frac{dU}{dt} = AU = UA$$

Note that the inverse of $U(t)$ is $U(-t)$. By using an operator norm, e.g. for matrices or bounded operators A, $\|A\| = \sup \|Av\|/\|v\|$, $v \neq 0$, these equations hold in the operator algebra. There are two further natural interpretations which we will see:

1) In terms of a representation, i.e., acting on a fixed vector v, we have the vector-valued function:

$$u(t) = e^{tA} v, \qquad u(0) = v$$

   And $du/dt = Au$. Conversely, for finite-dimensional spaces one can compute explicitly the solution to $du/dt = Au$, $u(0) = v$ as $\exp(tA)v$.

2) On an inner product space, (pre-)Hilbert space, one considers the matrix elements

$$u(t) = \langle e^{tA} v, w \rangle$$

   for fixed vectors v, w. Then $u(t)$ is a scalar function satisfying

$$\frac{du}{dt} = \langle A e^{tA} v, w \rangle, \qquad u(0) = \langle v, w \rangle$$

In our study, 2) will turn out to be always valid. This is discussed in detail in Chapter 3.

As for case 1), on a vector space, the operator A *acts nilpotently* if, for any given vector v, for all n sufficiently large, $A^n v = 0$. In such a case, both 1) and 2) make sense since the exponentials can be defined by finite series. Here we will be mainly interested in representations on polynomials. In that case, an operator A, such as $D = d/dx$, such that $\deg(Ap(x)) < \deg(p(x))$ for all polynomials $p(x)$, acts nilpotently.

Remark. In any case, we will not need a 'general theory'. For further information, see, e.g., Goldstein[24], Hille and Phillips[27].

Our principal interpretation of $\exp(tA)v$ is as the *generating function* for the sequence $\{A^n v\}_{n \geq 0}$:

$$\sum_{n=0}^{\infty} \frac{t^n}{n!} A^n v$$

In this case, it may be possible to give an operator formulation of e^{tA}, cf. section §IV below, in terms of operators with case 1) above understood. Another example arises when v is an eigenvector (eigenfunction) of A, $Av = \lambda v$ so that $e^{tA} v = e^{\lambda t} v$. In the present work, the formulas are interpreted as acting on representation spaces so that, in fact, $\exp(tA)v$ will be analytic functions of t and the interpretation as generating functions will be clear.

One can exponentiate Prop. 1.2.3, multiplying both sides by $t^n/n!$ and summing:

$$e^{t\Theta}(X) = e^{tA} X e^{tB}$$

For the case when $B = -A$ we have:

1.3.1 Proposition.

$$e^{tA} X e^{-tA} = e^{t \operatorname{ad} A} X$$

II.   Heisenberg-Weyl algebra

Recall Leibniz' rule: $(fg)' = f'g + fg'$, the primes denoting differentiation with respect to x. Introduce the notations:

$$D = \frac{d}{dx} \quad \text{and} \quad x = \text{multiplication by } x$$

i.e., $Df(x) = f'(x)$ and $xf(x) = xf(x)$. Then Leibniz' rule, replacing g by x, says that

$$D(xf(x)) = x(Df(x)) + (Dx)f(x)$$

With $(x)' = 1$, this yields the operator identity:

$$[D, x] = Dx - xD = 1$$

where 1 indicates multiplication by 1, i.e., the identity operator. We will see shortly that starting from this relation one can recover calculus for polynomials, exponentials, and eventually for more general functions via suitable limiting procedures. One method is via the theory of Fourier integrals. In Chapter 3, we will use Fock space constructions. We will be mainly concerned with polynomials, and with exponential functions, as in section 1.3, as generating functions for relations involving polynomials. We henceforth make the convention:

REMARK: In the following, an 'arbitrary function' — f, g, etc. — means a polynomial or a linear combination of exponentials with polynomial coefficients. By this convention all our formulas hold without further comment. One can extend much of the theory to formal power series. To deal with other interesting classes of functions requires some theory — see Chapter 3 — and our approach fits nicely with such extensions as they are needed.

Resuming our discussion, any two operators, which we conventionally denote by R and V such that $[V, R] = 1$ are said to generate the *Heisenberg-Weyl algebra*. For brevity we denote this: *HW algebra* .

Generally we do not insist that the commutator is the identity, just that it be a scalar operator. For example, with $d = hD$ for a scalar h, we have $[d, x] = h$. We thus have

2.1 Definition.   The *Heisenberg algebra* is the Lie algebra with basis $\{\, d, x, h \,\}$ satisfying

$$[d, x] = h, \qquad [d, h] = [x, h] = 0$$

Then the HW algebra is the associative algebra generated by $\{\, d, x, h \,\}$ with the Lie brackets realized as commutators. Note that the h actually indicates the operator of multiplication by the number h. From physics, we borrow the terminology *bosons* to denote specifically that we are taking $h = 1$.

2.1 COMMUTATION RULES

There are two basic techniques for calculating commutation rules. One way is to use induction and find the result for polynomials. The other is to use exponentials — generating functions. Here we start directly with polynomials.

2.1.1 Proposition.   For the HW algebra

$$[d, f(x)] = h f'(x)$$
$$[f(d), x] = h f'(d)$$

Proof: Use Prop. 1.2.2 to find, e.g.,

$$[d, x^n] = \sum x^{n-j-1} h\, x^j = h\, n x^{n-1}$$

For exponential functions, e.g. e^{ax}, multiply both sides by $a^n/n!$ and sum. Similarly for $[f(d), x]$.   ∎

We would like to get commutation formulas for polynomials in x and d. I.e., a formula of the type $d^n x^m = \sum c(n, m, j) x^{m-j} d^{n-j} h^j$. A general way to derive such a formula is to use generating functions. This we do in the next section.

2.2   EXPONENTIAL COMMUTATION RULES AND MATRIX ELEMENTS

The exponential commutation rules are in fact formulas for multiplying HW group elements. They are the essential formulas needed to compute the general group law (see §2.3). The problem in general is how to commute a function of d past a function of x.

2.2.1 Proposition.   *The HW exponential commutation rules:*

$$e^{td}\, e^{ax} = e^{ax}\, e^{td}\, e^{ath}$$

Proof: Apply Prop. 2.1.1 with $f(x) = e^{ax}$. Then we have $[d, e^{ax}] = ah\, e^{ax}$. Now write this as $(\operatorname{ad} d)\, e^{ax} = ah\, e^{ax}$ and apply Prop. 1.3.1 to get the result, after multiplying on the right by e^{td}.   ∎

Notice that this is an extension of the usual rule from calculus:

$$\frac{d}{dx}\, e^{ax} = a\, e^{ax} \tag{2.2.1}$$

i.e., the functions $\exp(ax)$ are eigenfunctions of d/dx. Thus,

$$e^{th\, d/dx}\, e^{ax} = e^{ath}\, e^{ax}$$

The difference here is that we are keeping track of the d as a partner variable with x. This extension to looking at both variables (x, d) is analogous to the idea in mechanics of considering the motion of a particle. You need both position and velocity to determine the motion. The space of both variables is called *phase space.*

Now, in Prop. 2.2.1 expand out in powers of t and a to find (recall the notation from the Introduction, §II):

2.2.2 Proposition.   *General Leibniz Rule:*

$$d^n x^m = \sum_j \binom{n, m}{j} x^{m-j} d^{n-j} h^j$$

2.2.1 Differential technique

Let us derive the exponential commutation rule in another way — using the *differential technique*. This will be useful later when dealing with more complicated cases. The idea is that differentiating the exponentials with respect to the appropriate parameters brings down the elements of the algebra as in eq. (2.2.1) (cf. §1.3). Suppose a rule of the form:

$$e^{td}\, e^{ax} = e^{ax}\, e^{V(t)d}\, e^{U(t)h}$$

Denote the common expression by E. Then differentiating with respect to t gives:

$$dE = EV'd + EU'h \qquad (2.2.1.1)$$

On the other hand the basic commutation rule $[d, e^{ax}] = ah\, e^{ax}$, Prop. 2.1.1, cf. eq. (2.2.1), gives the left hand side as $dE = Ed + Eah$. Comparing with eq. (2.2.1.1) yields:

$$V' = 1, \qquad U' = a$$

which, with the initial conditions $V(0) = U(0) = 0$, gives $V = t$, $U = at$ as in Prop. 2.2.1 above.

2.2.2 Associated calculus

So how does the usual calculus we learn fit into the picture? From the general Leibniz rules, Prop. 2.2.2, one recovers calculus on polynomials by considering a representation of the HW algebra. This is constructed as follows. Take a vector Ω such that $D\Omega = 0$. (Think of the constant function, 1.) Then build a vector space by applying the algebra to Ω. Since $D\Omega = 0$, the basis for the space consists of vectors $\psi_n = x^n\Omega$, which are identified with the usual functions x^n. The commutation rule $[D, x^n] = nx^{n-1}$ when applied to the vector Ω gives $Dx^n\Omega - x^nD\Omega = nx^{n-1}\Omega$, i.e., since $D\Omega = 0$,

$$D\, x^n\Omega = nx^{n-1}\Omega \qquad (2.2.2.1)$$

and we have recovered ordinary calculus. Since Ω is chosen so that it is annihilated by the operator D, the terminology used, from physics, is to call it a *vacuum vector* or *vacuum state*. In general, a single vector that generates the representation space by the action of the algebra is called a *cyclic vector*. For the algebra generated by d, x, and h, the vector Ω is chosen to satisy $d\Omega = 0$, $h\Omega = h\Omega$, where on the right hand side the h stands for the scalar h itself multiplying the vector Ω. Then we have $d\, x^n\Omega = h\, nx^{n-1}\Omega$, with the identification of d as the operator $h\, d/dx$.

For clarity, we present some conventions regarding our terminology.

2.2.2.1 Definition. The *standard HW algebra* is generated by V and R satisfying $[V, R] = 1$. We will also refer to standard generators V, R as *bosons* or a *boson pair*.

It is important to observe the following

2.2.2.2 Proposition.   *Let ψ_n be a basis for a vector space. Then a representation of the standard HW algebra is given by the operators V and R according to the action:*

$$R\psi_n = \psi_{n+1}, \qquad V\psi_n = n\psi_{n-1},$$

Note that $\psi_n = R^n\Omega$, $n \geq 0$, where we identify Ω as ψ_0. For the case of general $h > 0$, we have

2.2.2.3 Proposition.   *Let ψ_n be a basis for a vector space. Then a representation of the HW algebra is given by the operators L, R, and h, according to the action:*

$$R\psi_n = \psi_{n+1}, \qquad L\psi_n = h\,n\psi_{n-1}, \qquad h\psi_n = h\psi_n$$

Remark. Of course, we see that $L = hV$. The notation L, R, V corresponds to *lowering*, *raising*, and *velocity* operators respectively. The significance of the operator V as velocity will be seen in Chapter 5.

2.2.3 Matrix elements

We want to discuss as well the matrix elements of the group e^{td} acting on the basis $x^n\Omega$. Recall that if ψ_n is a basis in a vector space, then the matrix elements, M_{kn}, of an operator A are defined by the relations

$$A\psi_n = \sum_k M_{kn}\psi_k$$

Here we identify the basis elements ψ_n with $x^n\Omega$ of the preceding section.

2.2.3.1 Proposition.   *The matrix elements of the group e^{td} acting on the basis $x^n\Omega$ are given by*

$$M_{kn}(t) = \binom{n}{k} t^{n-k} h^{n-k}$$

Proof: This follows from the commutation rule $e^{td} x^n = (x + th)^n e^{td}$ applied to the vacuum state.   ∎

Remark. Notice that for each n the span of $\{\,\Omega, x\Omega, x^2\Omega, \ldots, x^n\Omega\,\}$ is an invariant subspace for the group. I.e., for each n we have a finite-dimensional representation of the group e^{td}, which is the group of translations generated by h. The group law

$$e^{td} e^{sd} = e^{(s+t)d}$$

applied to the basis $x^n\Omega$ yields the

2.2.3.2 Proposition.   *The matrix elements $M_{kn}(t)$ are a representation of the translation group, i.e.*

$$M_{kn}(s + t) = \sum_j M_{kj}(s)\, M_{jn}(t)$$

Below, §3.3.2, we will do similar calculations based on the sl(2) algebra. In Chapter 7 we will extend these calculations and will find interesting representations of the translation group as well.

2.3 GROUP LAW

From the Lie algebra, we form group elements by exponentiation. E.g., from the HW algebra we have

$$g(a,b,c) = e^{ax}\, e^{bd}\, e^{ch}$$

as group elements. Note that $g(0,0,0)$ is the identity. In general, one cannot form a group in the sense of permitting arbitrary values of the coordinates — the parameters (a,b,c), for example. But one always can multiply elements close to the identity. The details vary with the group, so for our purposes we make the

2.3.1 Definition.   The *group law* is the rule for multiplying two group elements that are sufficiently near the identity.

For the HW group, the parameters (a,b,c) can take any values in $\mathbf{R}^3$ or $\mathbf{C}^3$, e.g., corresponding to the *real* or *complex* HW group, respectively. I.e., there are no restrictions placed on the parameter values. Thus, the exponentials of the HW algebra form a group.

Prop. 2.2.1 yields:

2.3.2 Proposition.   *The HW group law:*

$$g(a,b,c)\,g(A,B,C) = g(a+A,b+B,c+C+bA)$$

The identity is $g(0,0,0)$, as noted above. The inverse of $g(a,b,c)$ is readily calculated to be $g(-a,-b,-c+ab)$. Note that as in §2.2 we can interpret the group law as the generating function for the general Leibniz rules. This will be our general point of view.

2.4 NUMBER OPERATOR AND OSCILLATOR ALGEBRA

On the basis $\psi_n = x^n\Omega$, in addition to the operators D and (multiplication by) x, we can include the operator xD, that acts according to

$$(xD)\, x^n\Omega = n\, x^n\Omega$$

i.e., $(xD)\psi_n = n\psi_n$. Thus, the basis vectors are all eigenvectors of xD, with corresponding eigenvalues $0,1,2,\ldots$. We have:

2.4.1 Definition.   Let ψ_n be a basis for a vector space. The operator ν such that $\nu\psi_n = n\psi_n$ is called the *number operator* .

The terminology is from the physics of bosons, where n is the number of particles; see §5.3 for more details concerning the harmonic oscillator (hence the name of the algebra), where the energy levels correspond to the number of photons.

The Lie algebra with the operators x, D, xD, and 1 as basis is referred to as the *oscillator algebra* .

Remark. In terms of our general notations, with bosons V, R, cf. Def. 2.2.2.1, we have as a basis for a standard oscillator algebra: R, V, RV, and the identity. We can include the scalar h by setting $L = hV$, using as basis: R, L, RV, and h. Cf. Prop. 2.2.2.3. These act on the basis $\psi_n = R^n \Omega$.

Via the HW theory we have

2.4.2 Proposition. *Commutation rules for the oscillator algebra:*

$$[xD, f(x)] = xf'(x), \qquad\qquad [f(D), xD] = Df'(D)$$

$$f(xD)\, x = x\, f(xD + 1), \qquad\qquad D\, f(xD) = f(xD + 1)\, D$$

$$a^{xD} f(x) = f(ax)\, a^{xD}, \qquad\qquad f(D)\, a^{xD} = a^{xD} f(aD)$$

Proof: The first line follows directly from HW relations. For the second line, iterate the relation $xD\,x = x\,(xD + 1)$ in the associative algebra, to get $(xD)^n x = x\,(xD + 1)^n$. Commuting D past $f(xD)$ is similar. For the third line, use the result of the second line with $f(xD) = a^{xD}$ to get $a^{xD} x = x\, aa^{xD}$. Iterate this to get $a^{xD} x^n = x^n a^n a^{xD} = (ax)^n a^{xD}$. Similarly, conclude the result for $f(D)$. ■

Remark. Applying the oscillator algebra to the vacuum Ω gives the representation space with functions of x in the form $f(x)\Omega$, as for the HW algebra. The operator xD generates the *group of dilations* $a^{xD} f(x)\Omega = f(ax)\Omega$. This explains the use of the notation a^{xD} rather than the exponential notation e^{txD}.

III. sl(2) algebra

Now we introduce another of our protagonists. This algebra arises immediately as soon as one looks for matrix realizations of Lie algebras. The notation *sl* means *special linear* the term special denoting matrices of determinant one. The *group* of $n \times n$ matrices of determinant one is denoted SL(n), or, if one wants to indicate specifically that it is over the reals, e.g., then SL(n, **R**). The corresponding Lie algebra is denoted by lower case letters: sl(n). From the point of view of the Lie algebra, given a group element of the form $g(t) = e^{tA}$ with $\det g(t) = 1$, the relation $\det e^A = e^{\operatorname{tr} A}$ translates into $\operatorname{tr} A = 0$. Thus, *sl(n)* denotes the Lie algebra of $n \times n$ matrices of trace zero. If A and B are $n \times n$ matrices, then $\operatorname{tr}(AB) = \operatorname{tr}(BA)$ is the same as $\operatorname{tr}([A, B]) = 0$. That is, commutators are always in sl(n).

So, *sl(2)* denotes the Lie algebra of 2×2 matrices of trace zero. This is a 3-dimensional vector space, and we have the standard basis:

$$\Delta = \begin{pmatrix} 0 & 0 \\ -1 & 0 \end{pmatrix}, \qquad R = \begin{pmatrix} 0 & 1 \\ 0 & 0 \end{pmatrix}, \qquad \rho = \begin{pmatrix} 1 & 0 \\ 0 & -1 \end{pmatrix} \qquad (3.1)$$

3.1 STANDARD FORM FOR sl(2)

One easily checks the commutation relations for the basis, eq. (3.1).

3.1.1 Proposition. *The basis Δ, R, ρ satisfies the commutation relations:*

$$[\Delta, R] = \rho, \qquad [\rho, R] = 2R, \qquad [\Delta, \rho] = 2\Delta$$

3.1.2 Definition. Any three operators satisfying the commutation relations of Prop. 3.1.1 will be referred to as a basis for the *standard sl(2) algebra* or as a standard basis.

Remark. Observe that in the associative algebra, with brackets realized as commutators,

$$\rho R = R(\rho + 2), \qquad \Delta \rho = (\rho + 2)\Delta \qquad (3.1.1)$$

That is, ρ acts like $2\times$ number operator, cf. Def. 2.4.1 and Prop. 2.4.2.

The notation comes from the Laplacian for $\mathbf{R}^N$ acting on radial functions. For $\mathbf{R}^2$, we have

3.1.3 Proposition. *A realization of the standard sl(2) Lie algebra is given by:*

$$\Delta = \tfrac{1}{2}\left(\frac{\partial^2}{\partial x^2} + \frac{\partial^2}{\partial y^2}\right), \quad R = \tfrac{1}{2}(x^2 + y^2), \quad \rho = x\frac{\partial}{\partial x} + y\frac{\partial}{\partial y} + 1$$

Proof: Using the HW commutation rules, e.g., one has $[D^2/2, x^2/2] = xD + \tfrac{1}{2}$, $[xD, x^2/2] = x \cdot x = x^2$, etc. ∎

Remark. Note that if we have the Laplacian in N dimensions, then the commutation relations hold with the dimension reflected in the expression for ρ, namely,

$$\Delta = \tfrac{1}{2}\sum \frac{\partial^2}{\partial x_j^2}, \quad R = \tfrac{1}{2}\sum x_j^2, \quad \rho = \sum x_j\frac{\partial}{\partial x_j} + \frac{N}{2} \qquad (3.1.2)$$

3.2 COMMUTATION RULES FOR sl(2)

To develop the calculus associated with the sl(2) algebra, we want to find the commutation rules in the associative algebra, corresponding to any representation of sl(2).

3.2.1 Proposition.   *For the standard sl(2) algebra:*

$$[f(\Delta), R] = \rho f'(\Delta) + \Delta f''(\Delta)$$

$$[\Delta, f(R)] = f'(R)\rho + R f''(R)$$

$$[f(\Delta), \rho] = 2\Delta f'(\Delta)$$

$$[\rho, f(R)] = 2R f'(R)$$

Proof: We prove the first and last lines, as the middle two follow similarly. First, consider $[\rho, f(R)]$, recalling eq. (3.1.1). We have, iterating $\rho R = R(\rho + 2)$:

$$\rho R^n = R^n(\rho + 2n) = R^n \rho + 2n R^n$$

That is,

$$[\rho, R^n] = 2R (R^n)'$$

This extends to polynomials; and to exponentials e^{ax} by multiplying by $a^n/n!$ and summing. For $[f(\Delta), R]$, use Prop. 1.2.2, in the sum replacing j by $n - 1 - j$. Thus, using eq. (3.1.1) in the third line:

$$\begin{aligned}
\Delta^n R - R\Delta^n &= \sum \Delta^j [\Delta, R] \Delta^{n-1-j} \\
&= \sum \Delta^j \rho \Delta^{n-1-j} \\
&= \sum (\rho + 2j) \Delta^j \Delta^{n-1-j} \\
&= \rho n \Delta^{n-1} + n(n-1)\Delta^{n-1} \\
&= \rho (\Delta^n)' + \Delta (\Delta^n)''
\end{aligned}$$

which extends to general f.   ■

3.3 EXPONENTIAL COMMUTATION RULES AND MATRIX ELEMENTS

First we see how to commute functions of ρ past functions of Δ and of R. Comparing eq. (3.1.1) and Prop. 2.4.2 shows that a^ρ acts as dilation by a^2, i.e.

3.3.1 Proposition.

$$a^\rho f(R) = f(Ra^2) a^\rho, \qquad f(\Delta) a^\rho = a^\rho f(a^2 \Delta)$$

For $e^{t\Delta} e^{aR}$ we use the differential technique as in §2.2.1.

3.3.2 Proposition.   *For the standard sl(2) algebra*

$$e^{t\Delta}\, e^{aR} = \exp\left(\frac{aR}{1-at}\right)(1-at)^{-\rho}\exp\left(\frac{t\Delta}{1-at}\right)$$

Proof: Suppose we have the form

$$E = e^{t\Delta}\, e^{aR} = e^{V(t)R}\, e^{W(t)\rho}\, e^{U(t)\Delta} \tag{3.3.1}$$

Differentiating with respect to t:

$$\frac{\partial E}{\partial t} = \Delta E = V'(t)RE + e^{VR}\,\rho W'(t)e^{W\rho}\,e^{U\Delta} + EU'(t)\Delta \tag{3.3.2}$$

(in the exponents the variable t is understood)

On the right side of eq. (3.3.1), we use the commutation rules, Propositions 3.2.1 and 3.3.1:

$$[\Delta, e^{VR}] = V(t)e^{VR}\rho + RV(t)^2 e^{VR} \qquad \text{and} \qquad \Delta e^{W\rho} = e^{2W}\, e^{W\rho}\,\Delta$$

This gives

$$\Delta E = V(t)e^{VR}\rho e^{W\rho}\,e^{U\Delta} + RV(t)^2 E + e^{2W}\, E\Delta$$

Comparing with the coefficients in eq. (3.3.2) yields the system:

$$V' = V^2, \qquad W' = V, \qquad U' = e^{2W}$$

with initial conditions $V(0) = a$, $W(0) = U(0) = 0$. These are readily solved to give the result.   ■

We can now give the general commutation rule for sl(2). From the corresponding classical result for the HW algebra, we refer to such a rule as the *general Leibniz rule* for the algebra under consideration.

3.3.3 Proposition.   *The general Leibniz rule for sl(2) is*

$$\Delta^n R^m = \sum_j \binom{m,n}{j} R^{m-j}\Delta^{n-j}(\rho + m - n)_j$$

Proof: In Prop. 3.3.2, first pull the middle term to the right, using Prop. 3.3.1:

$$\exp\left(\frac{aR}{1-at}\right)(1-at)^{-\rho}\exp\left(\frac{t\Delta}{1-at}\right) = \exp\left(\frac{aR}{1-at}\right)\exp(t(1-at)\Delta)(1-at)^{-\rho}$$

Now expand (summation convention used on the right):

$$e^{t\Delta}\, e^{aR} = \sum\sum \frac{t^n a^m}{n!\,m!}\Delta^n R^m = \frac{a^\lambda t^\mu}{\lambda!\,\mu!} R^\lambda \Delta^\mu (1-at)^{-\rho-\lambda+\mu}$$

$$= \sum_j \frac{a^{\lambda+j} t^{\mu+j}}{\lambda!\,\mu!} R^\lambda \Delta^\mu \frac{(\rho+\lambda-\mu)_j}{j!}$$

Let $\lambda + j = m$, $\mu + j = n$ and compare coefficients.   ■

3.3.1 Calculus

Here to get a calculus we choose a vacuum state Ω that satisfies

$$\Delta\Omega = 0, \qquad \rho\Omega = c\Omega \tag{3.3.1.1}$$

where c is a scalar. Applying the algebra to the vector Ω yields a vector space with basis $\psi_n = R^n\Omega$. The commutation rules, Prop. 3.2.1, yield

3.3.1.1 Proposition.   *The sl(2) calculus is given by the action of Δ on the basis $R^n\Omega$:*

$$\Delta\, R^n\Omega = n(c + n - 1)R^{n-1}\Omega$$

Proof: Apply the commutation rule $[\Delta, R^n] = nR^{n-1}\rho + n(n-1)R^{n-1}$ to Ω.
∎

Similarly, we find that ρ acts as a modified number operator:

3.3.1.2 Proposition.   *The action of ρ on the basis $R^n\Omega$ is given by*

$$\rho\, R^n\Omega = (c + 2n)R^n\Omega$$

Proof: Use the rule $[\rho, f(R)] = 2Rf'(R)$ applied to Ω.   ∎

3.3.2 Matrix elements

As in the HW case, for each n, we have a finite-dimensional representation of the action of the group generated by Δ.

3.3.2.1 Proposition.   *The matrix elements of the group $e^{t\Delta}$ acting on the basis $R^n\Omega$ are given by*

$$M_{kn}(t) = \binom{n}{k}\binom{n + c - 1}{n - k}(n - k)!\, t^{n-k}$$

Proof: The action of Δ according to Prop. 3.3.1.1 iterated k times yields

$$\Delta^k R^n\Omega = n^{(k)}(n + c - 1)^{(k)} R^{n-k}\Omega$$

Multiplying by $t^k/k!$ and summing yields the result, after replacing k by $n - k$.
∎

Here it is still a question of the translation group, since we are just considering the group generated by Δ.

3.4 GROUP LAW

Using Propositions 3.3.1 and 3.3.2, we can calculate the group law for the sl(2) algebra. Define the group elements

$$g(a, b, c) = e^{aR} b^\rho e^{c\Delta}$$

Here the identity is $g(0, 1, 0)$.

3.4.1 Proposition. *The group elements satisfy*

$$g(a, b, c)\, g(A, B, C) = g\left(a + \frac{Ab^2}{1 - Ac}, \frac{bB}{1 - Ac}, C + \frac{cB^2}{1 - Ac}\right)$$

Proof: By Prop. 3.3.2

$$g(a, b, c)\, g(A, B, C) = g(a, b, 0)\, g\left(\frac{A}{1 - Ac}, \frac{1}{1 - Ac}, \frac{c}{1 - Ac}\right) g(0, B, C)$$

Applying Prop. 3.3.1 to pull the b^ρ and B^ρ terms through gives

$$g(a, 0, 0)\, g\left(\frac{Ab^2}{1 - Ac}, \frac{bB}{1 - Ac}, \frac{cB^2}{1 - Ac}\right) g(0, 0, C)$$

which yields directly the required result. ■

Remark. Note here the requirement for the group elements to be near the identity (Def. 2.3.1). Starting from the identity, the parameters cannot extend past $|Ac| < 1$.

3.5 SCALED VERSION OF sl(2)

As in the HW case, it is natural to consider a scaled version of the algebra. I.e., from $[D, x] = 1$, we can scale D by h to $d = hD$ getting $[d, x] = h$. Similarly, here we can scale $\Delta \to \beta\Delta$. We use this occasion to introduce our conventional notations.

3.5.1 Proposition. *Let $L = \beta\Delta$. Then, redefining ρ directly in terms of L and R, one has the commutation relations*

$$[L, R] = \rho, \qquad [\rho, R] = 2\beta R, \qquad [L, \rho] = 2\beta L$$

We give a realization in terms of bosons V, R. The following is immediate from the HW algebra commutation rules.

3.5.2 Proposition.   *Let $[V, R] = 1$ be standard boson operators. Then*

$$L = cV + \beta RV^2, \qquad R = R, \qquad \rho = c + 2\beta RV$$

satisfy the scaled sl(2) commutation relations.

E.g., $L = xD^2$, $\rho = 2xD$, $R = x$ is a realization of the standard (i.e., $\beta = 1$) sl(2) algebra. And we observe the general feature, cf. Prop. 2.2.2.3:

3.5.3 Proposition.   *Let ψ_n be a basis for a vector space. Then, for given scalars c and β, a representation of the scaled sl(2) algebra is given by the operators R, L, and ρ according to the action:*

$$R\psi_n = \psi_{n+1}, \qquad L\psi_n = n(c + \beta(n-1))\psi_{n-1}, \qquad \rho\psi_n = (c + 2n\beta)\psi_n$$

IV.   Splitting formulas

When exponentiating elements of the Lie algebra to form group elements, it is natural to take an arbitrary element of the Lie algebra and generate a group element. E.g., for the HW algebra, take $X = \alpha x + \beta D + \gamma h$, and form the group element exp(X). The parameters (α, β, γ) of group elements formed this way are called *coordinates of the first kind* , while the coordinates we have used above for group elements are referred to as *coordinates of the second kind* . Both types of coordinates will be important for our work. Here we use the differential technique to derive the basic formulas relating these two coordinate systems (near the identity of the group). Since the relation is one of factoring the exponential of a given element, we call formulas of this type *splitting formulas* .

For each group, we take an element X from the algebra and compute in terms of coordinates of the second kind the *one-parameter subgroup* generated by X, i.e., the group elements of the form exp(sX).

4.1   HEISENBERG-WEYL GROUP

We consider the HW algebra with basis x, d, h.

4.1.1 Proposition.   *For the HW group we have*

$$e^{s(x+ad+bh)} = e^{sx}\,e^{sad}\,e^{(sb+s^2a/2)h}$$

Proof: Observe that the result for general h follows from the result for $h = 1$ by replacing a by ah. So we consider $\exp(s(x + aD))$ and start with the form, denoting this by E:

$$E = e^{s(x+aD)} = e^{V(s)x} e^{U(s)D} e^{W(s)} \tag{4.1.1}$$

Now differentiating with respect to s brings down the multiplier $x + aD$. We first consider: $(x + aD)E = xE + a(DE)$. For the term DE, we have to pull the D past $\exp(V(s)x)$. The HW commutation rules give:

$$D\, e^{V(s)x} = e^{V(s)x} (D + V(s))$$

Thus,

$$\frac{\partial E}{\partial s} = (x + aD)E = xE + aE(D + V(s))$$

while on the right side of eq. (4.1.1), differentiating brings down one factor at a time:

$$\frac{\partial E}{\partial s} = V'(s)xE + EU'(s)D + EW'(s)$$

Now equate corresponding coefficients of x, D, 1 to get the system:

$$V'(s) = 1, \qquad U'(s) = a, \qquad W'(s) = aV(s)$$

with the initial conditions $V(0) = U(0) = W(0) = 0$ at the identity. These are readily solved to find the result stated. ∎

Remark. Observe that for $b = 0$, $h = 1$, this may be written in the form

$$e^{V(s)x} e^{aH(s)} e^{V(s)aD}$$

with $H(s) = s^2/2$ and $V(s) = H'(s) = s$.

4.2 OSCILLATOR GROUP

The oscillator group is generated by the oscillator algebra of §2.4. That is, we adjoin the number operator to the HW basis, taking the basis in the standard form x, xD, D, and 1.

4.2.1 Proposition. *For the oscillator group:*

$$e^{s(x+a\,xD+bD)} = e^{bH(s)} e^{V(s)x} e^{as\,xD} e^{V(s)bD}$$

where $H(s) = (e^{as} - 1 - as)/a^2$ and $V(s) = H'(s) = (e^{as} - 1)/a$.

Proof: Start with the form

$$E = e^{s(x+axD+bD)} = e^{V(s)x} e^{W(s)xD} e^{U(s)D} e^{T(s)}$$

Here we use Prop. 2.4.2 to pull the terms xD and D through. We find, using the commutation rules $(xD)f(x) = f(x)(xD) + xf'(x)$ and $D f(xD) = f(xD+1)D$:

$$\frac{\partial E}{\partial s} = xE + a(xV(s)E + e^{Vx} xD e^{WxD} e^{UD} e^{T}) + b(V(s)E + e^{W(s)} ED)$$

(where, as usual, in the exponents we generally suppress the s variable)
Comparing this with

$$V'(s)xE + e^{Vx} W'(s)xD e^{WxD} e^{UD} e^{T} + e^{Vx} e^{WxD} U'(s)D e^{UD} e^{T} + ET'(s)$$

yields the system, with the variable s understood:

$$V' = 1 + aV, \qquad W' = a, \qquad U' = be^{W}, \qquad T' = bV$$

with the usual zero initial conditions. First solve for $V(s) = (\exp(as) - 1)/a$. Then integrate to get T. Similarly, integrate immediately to get W, and then U.   ∎

4.3 SL(2) GROUP

We take the standard basis Δ, R, ρ.

4.3.1 Proposition.   *For the SL(2) group,*

$$e^{s(R+a\rho+b\Delta)} = e^{V(s)R} e^{\rho H(s)} e^{V(s)b\Delta}$$

where, with $\delta^2 = a^2 - b$,

$$H(s) = \log\left(\frac{\delta \operatorname{sech} \delta s}{\delta - a\tanh \delta s}\right) \qquad \text{and} \qquad V(s) = b^{-1}(H'(s) - a) = \frac{\tanh \delta s}{\delta - a\tanh \delta s}$$

Proof: As in the above proofs, start with the form

$$E = e^{s(R+a\rho+b\Delta)} = e^{V(s)R} e^{H(s)\rho} e^{U(s)\Delta} \tag{4.3.1}$$

Differentiating with respect to s yields, using the commutation rules, Propositions 3.2.1 and 3.3.1:

$$\frac{\partial E}{\partial s} = RE + a(e^{VR} \rho e^{H\rho} e^{U\Delta} + 2V(s)RE)$$

$$+ b(RV(s)^2 E + V(s)e^{VR} \rho e^{H\rho} e^{U\Delta} + e^{2H(s)} E\Delta)$$

Comparing with the result of differentiating the right hand side of eq. (4.3.1), one arrives at the system:

$$V' = 1 + 2aV + bV^2, \qquad H' = a + bV, \qquad U' = be^{2H}$$

The equation for V may be directly integrated, calculating:

$$\int \frac{dv}{1 + 2av + bv^2}$$

(One may also verify directly that $V(s)$ as quoted above satisfies the differential equation.) Differentiating $V' = 1 + 2aV + bV^2$ yields $V'' = 2(a + bV)V'$. But with $H' = a + bV$, this gives $V'' = 2H'V'$, which integrates to

$$V' = e^{2H} \qquad (4.3.2)$$

Hence, $U' = bV'$ so that $U = bV$. We get via direct calculation of V'

$$e^H = \sqrt{V'} = \frac{\delta \operatorname{sech} \delta s}{\delta - a \tanh \delta s}$$

■

And this completes our calculation of the splitting formulas for these groups.

Remark. Note the occurrence of the *Riccati equation* — a first order *quadratic* differential equation — for the function $V(s)$. In fact, the Riccati equation, hidden, occurs for the HW and oscillator cases as well. For the HW case, one has simply $a = b = 0$, while for the oscillator group, $b = 0$, and a replaces $2a$. We will see eventually the deeper significance of these facts — in Chapters 5 and 6.

Some references for Lie algebras and Lie groups to note:

1. For a primarily mathematical point of view, see Bourbaki[9], Fulton&Harris[21], Helgason[25], and Gilmore[23], who writes for physicists as well.

2. The works Bäuerle&de Kerf[4], Belinfante&Kolman[5], Klimyk&Vilenkin[30] are aimed for applications as well as providing theoretical material. Belinfante&Kolman includes material concerning computational aspects of Lie algebras.

3. For connections with special functions the book of Vilenkin[46] is now a classic, and see as well Klimyk&Vilenkin[30].

4. For the theory of angular momentum in physics, see Biedenharn&Louck[6].

V.   Exercises and examples

5.1 EXERCISES

1. Verify that the Jacobi identity holds for the Lie bracket given by the commutator, $[A, B] = AB - BA$.

2. a. Show that if $[A, B] = 0$, i.e., $AB = BA$, in an associative algebra, then
   $[A, CB] = [A, C]B$.

   b. Show generally that (in an associative algebra)

   $$[A, BC] = [A, B]C + B[A, C]$$

3. Prove the four Propositions on page 10.

4. Verify Proposition 1.3.1 for $A = \begin{pmatrix} 0 & 0 \\ \lambda & 0 \end{pmatrix}$, $X = \begin{pmatrix} a & b \\ c & d \end{pmatrix}$

5. Extend Proposition 2.2.2 to 'general functions' f, g:

   $$g(d)f(x) = \sum_{k=0}^{\infty} \frac{h^k}{k!} f^{(k)}(x) g^{(k)}(d)$$

6. From Proposition 2.3.2, find a formula for the nth power of $g(a, b, c)$. E.g.,
   $g(a, b, c)^2 = g(2a, 2b, 2c + ab)$.

7. Check the formula $g(a, b, c)^{-1} = g(-a, -b, -c + ab)$.

8. Verify the commutation rules in Proposition 2.4.2. Then do:

   a. $[(xD)^n, x]$

   b. $[D, (xD)^n]$

9. Define oscillator group elements

   $$g(p, q, r, s) = e^{px}\, e^{qxD}\, e^{rD}\, e^s$$

   for the basis x, xD, D, 1 of the oscillator algebra. Find the corresponding
   group law.

10. Prove Proposition 3.1.1.

11. Verify the claims that the operators given in Proposition 3.1.3 and in equation
    (3.1.2) are realizations of the sl(2) algebra.

12. a. Prove the two middle relations in Proposition 3.2.1

    b. Find $[\Delta^n, \rho]$, $[\Delta, R^n]$.

13. Finish the proof of Proposition 3.3.1.2.

14. a. Find the inverse, $g(a, b, c)^{-1}$, for the typical SL(2) group element in §3.4.

    b. Find a formula for $g(a, b, c)^2$ and $g(a, b, c)^3$.

    c. Try to find the general formula for $g(a, b, c)^n$.

15. a. Verify Proposition 3.5.1

    b. Find $[L, R^n]$ for the scaled sl(2) algebra.

    c. Verify Proposition 3.5.3.

16. Verify the formulas in §3.3 using the 2×2 matrix realization of sl(2) given in equation (3.1).

17. Verify the SL(2) splitting formula for the realization as 2×2 matrices.

5.2 MATRIX REALIZATION OF THE HW ALGEBRA AND GROUP

A realization of the HW algebra is given by the 3×3 matrices

$$
x = \begin{pmatrix} 0 & 0 & 0 \\ 0 & 0 & 1 \\ 0 & 0 & 0 \end{pmatrix}, \qquad
d = \begin{pmatrix} 0 & 1 & 0 \\ 0 & 0 & 0 \\ 0 & 0 & 0 \end{pmatrix}, \qquad
h = \begin{pmatrix} 0 & 0 & 1 \\ 0 & 0 & 0 \\ 0 & 0 & 0 \end{pmatrix}
$$

1. Verify the HW commutation relations.

2. Calculate the group elements $g(a, b, c)$ in matrix form.

3. Verify the group law, Proposition 2.3.2.

4. Verify the splitting formula, Proposition 4.1.1, in this matrix realization.

5.3 QUANTUM HARMONIC OSCILLATOR

The HW and oscillator algebras arise in the quantization of the harmonic oscillator, see Landau&Lifshitz[31] for background. The problem is to find the eigenvalues and eigenfunctions of the Schrödinger operator. For the harmonic oscillator we have the equation for the wave function ψ:

$$\tfrac{1}{2}\psi'' + \left(\frac{E}{\hbar\omega} - \frac{x^2}{2} \right) \psi = 0$$

i.e., we want eigenvalues and eigenfunctions of the operator $\frac{1}{2}(D^2 - x^2)$.

Let $R = (x - D)/\sqrt{2}$, $V = (x + D)/\sqrt{2}$.

1. Show that $[V, R] = 1$.

2. Verify that $\Omega = \exp(-x^2/2)$ is a vacuum state, i.e., $V\Omega = 0$. Thus, $\psi_n = R^n\Omega$ is a basis for the corresponding HW representation.

3. Show that the number operator $RV = \frac{1}{2}(x^2 - D^2) - \frac{1}{2}$.

4. Since $RV\psi_n = n\psi_n$, we have the relation

$$\tfrac{1}{2}(x^2 - D^2)\psi_n = (n + \tfrac{1}{2})\psi_n$$

with the spectrum $\{ n + \frac{1}{2} \}_{n \geq 0}$.

5. Calculate ψ_2, ψ_3, ψ_4 explicitly as functions of x. E.g.,

$$\psi_1(x) = R\Omega = \tfrac{1}{\sqrt{2}}(x + x)e^{-x^2/2} = \sqrt{2}\, x e^{-x^2/2}$$

5.4 VACUUMS AND FURTHER CONSTRUCTIONS FOR HW REPRESENTATIONS

We will see that the condition for a vacuum state is naturally formulated in terms of the number operator xD. We say that a function f is *homogeneous of degree p* if $f(\lambda x) = \lambda^p f(x)$, $\forall \lambda > 0$.

1. Show that f is homogeneous of degree p if and only if $f(x) = |x|^p g(x)$ where g is homogeneous of degree zero if and only if $f(x)$ satisfies $xDf = pf$ (cf. §2.4).

2. Show that $\psi_n = x^n\Omega$, $n \geq 0$, gives a basis for an HW representation: $x\psi_n = \psi_{n+1}$, $D\psi_n = n\psi_{n-1}$, $n \geq 1$.

3. Besides $\Omega = 1$, we have $\Omega_+ = \chi_{[0,\infty)}$, equal to 1 for $x \geq 0$, 0 for $x < 0$, as a vacuum in one dimension.

4. Furthermore, one can construct representations (Feinsilver[15]) by

   a. Using negative powers of x

   b. Differentiating Ω_+

5. Discuss vacuums for dimensions $N = 2$, $N = 3$, $N > 3$ (see below).

5.5 RADIAL AND HARMONIC FUNCTIONS

Let $f: \mathbf{R}^2 \to \mathbf{R}$. We will use (x, y) and (r, θ) to denote rectangular and polar coordinates on $\mathbf{R}^2$ respectively.

As in the previous section, a function f on $\mathbf{R}^2$ is homogeneous of degree p if $f(\lambda x, \lambda y) = \lambda^p f(x, y)$, $\forall \lambda > 0$.

1. Show that, acting on f,

$$r \frac{\partial}{\partial r} = x \frac{\partial}{\partial x} + y \frac{\partial}{\partial y}$$

$$\frac{\partial}{\partial \theta} = x \frac{\partial}{\partial y} - y \frac{\partial}{\partial x}$$

Check these directly for $r^2 = x^2 + y^2$, $\cos \theta = x / \sqrt{x^2 + y^2}$.

2. Using results of Problem 1,

   a. Verify that $[\frac{\partial}{\partial r}, \frac{\partial}{\partial \theta}] = 0$, using HW commutation rules.

   b. Check the formula for the Laplacian in polar coordinates:

$$\frac{\partial^2}{\partial x^2} + \frac{\partial^2}{\partial y^2} = r^{-2} \left(\left(r \frac{\partial}{\partial r} \right)^2 + \frac{\partial^2}{\partial \theta^2} \right)$$

$$= \frac{\partial^2}{\partial r^2} + \frac{1}{r} \frac{\partial}{\partial r} + \frac{1}{r^2} \frac{\partial^2}{\partial \theta^2}$$

3. a. Verify that f is homogeneous of degree p if and only if f is of the form $r^p g(\theta)$, where g, a function of θ only, is homogeneous of degree 0.

   b. Show that f (smooth) is homogeneous of degree p if and only if f satisfies *Euler's equation for homogeneous functions*

$$x \frac{\partial f}{\partial x} + y \frac{\partial f}{\partial y} = pf$$

4. For the sl(2) realization of Proposition 3.1.3, we have $\Delta = $ half the Laplacian and $\rho = r (\partial / \partial r) + 1$.

   a. Show that (§3.3.1) a vacuum Ω is a harmonic function that is homogeneous of degree zero. I.e., it is a function on the circle that is harmonic.

   b. Find two (linearly independent) solutions to the equations for the vacuum.

5. Do a similar study for $N = 3$ dimensions with

$$r \frac{\partial}{\partial r} = x \frac{\partial}{\partial x} + y \frac{\partial}{\partial y} + z \frac{\partial}{\partial z}$$

and, for rotations by angles θ_x, θ_y, θ_z about the x, y, z axes respectively, we have, e.g.,

$$\frac{\partial}{\partial\theta_x} = y\frac{\partial}{\partial z} - z\frac{\partial}{\partial y}$$

and so on. In particular,

a. Verify that

$$\frac{\partial^2}{\partial x^2} + \frac{\partial^2}{\partial y^2} + \frac{\partial^2}{\partial z^2} - \left(\frac{\partial^2}{\partial r^2} + \frac{2}{r}\frac{\partial}{\partial r}\right)$$

annihilates radial functions, i.e., functions depending on r only.

b. Use HW commutation rules and the formulas for angular derivatives to find an expression for r^2 times the operator in part a., and hence deduce a corresponding expression for the Laplacian.

6. Extend to $N > 3$.

5.6 ALGEBRAIC APPROACH TO INTEGRAL CALCULUS

The study of representations of the HW algebra gives an algebraic approach to differential calculus. It is natural to wonder about integral calculus. Here we present an approach based on the commutation relations

$$[J, x] = -J^2$$

which comes formally from the HW rule $[f(D), x] = f'(D)$ for $J = 1/D$. Let us see some consequences of this commutation relation.

1. Show that $[x, J^n] = nJ^{n+1}$. (Thus, x may be realized as the operator $J^2\,(d/dJ)$ on the representation space with basis $J^n\Omega$, where $x\Omega = 0$.)

2. Deduce that

$$e^{ax}\,Je^{-ax} = J + aJ^2 + \cdots = \frac{J}{1 - aJ}$$

And hence the commutation rules

a. $e^{ax}\,f(J) = f(J/(1 - aJ))e^{ax}$

b. $f(J)e^{ax} = e^{ax}\,f(J/(1 + aJ))$

3. With $f(J) = e^{bJ}$ in the previous problem, expand to find the general Leibniz rules (cf. Problem 5 of §5.1):

$$J^n x^m = \sum_k \binom{m, -n}{k} x^{m-k} J^{n+k}$$

and

$$g(J)f(x) = \sum_{k=0}^{\infty} \frac{1}{k!} f^{(k)}(x) J^{k+1} \left(-\frac{d}{dJ}\right)^k J^{k-1} g(J)$$

4. Deduce $[J, f(x)] = \sum_k (-1)^k f^{(k)}(x) J^{k+1}$.

5. a. The action of J as an operator is given by $Jf(x)\Omega = f(x)J\Omega + [J, f(x)]\Omega$, then dropping the Ω's, leaving functions of x. For the representation with $J^n\Omega = x^n/n!$, the result of the previous problem thus yields

$$Jf(x) = \int_0^x f(y)\,dy$$

(Try $f(x) = x^m$.)

b. For the representation induced from $Je^{x/a} = a\,e^{x/a}$, $a > 0$, i.e., $\Omega = e^{x/a}$, we find

$$Jf(x) = \int_{-\infty}^x f(y)\,dy$$

5.7 ADJOINT REPRESENTATION

As in an associative algebra we have the left and right multiplication operators (§1.1), for any Lie algebra we have the *adjoint representation* mapping $X \rightarrow \text{ad}(X)$, taking the Lie algebra into linear mappings on the algebra:

$$(\text{ad}\,X)(Y) = [X, Y]$$

(see comment after Proposition 1.2.3, and see Proposition 1.3.1)

1. Verify that this is a representation:

$$[\text{ad}\,X, \text{ad}\,Y] = \text{ad}\,[X, Y]$$

E.g., $(\text{ad}\,X)(\text{ad}\,Y)(Z) = [X, [Y, Z]]$.

2. Given a basis for the Lie algebra, we can thus find matrices satisfying the commutation relations of the Lie algebra. If $\{\xi_1, \ldots, \xi_N\}$ is a basis, then we define the coefficients c_{ij}^k via

$$(\text{ad}\,\xi_k)(\xi_j) = [\xi_k, \xi_j] = \sum_i c_{ij}^k \xi_i$$

i.e., the matrix representation of $\operatorname{ad}\xi_k$ is $(\operatorname{ad}\xi_k)_{ij} = c_{ij}^k$. The coefficients c_{ij}^k are called the *structure constants* of the Lie algebra. Verify that for the HW algebra with basis $\{\,x, d, h\,\}$, we have the adjoint representation:

$$
\operatorname{ad}x = \begin{pmatrix} 0 & 0 & 0 \\ 0 & 0 & 0 \\ 0 & -1 & 0 \end{pmatrix}, \qquad \operatorname{ad}d = \begin{pmatrix} 0 & 0 & 0 \\ 0 & 0 & 0 \\ 1 & 0 & 0 \end{pmatrix}, \qquad \operatorname{ad}h = \begin{pmatrix} 0 & 0 & 0 \\ 0 & 0 & 0 \\ 0 & 0 & 0 \end{pmatrix}
$$

Note that $\operatorname{ad}h$ equals the zero matrix, since h commutes with x, d.

3. Find (matrices for) the adjoint representation for the oscillator algebra, with basis $\{\,x, xd, d, h\,\}$, and sl(2), with basis $\{\,R, \rho, \Delta\,\}$.

5.8 NILPOTENT LIE ALGEBRAS

A *nilpotent Lie algebra* means that there exists n such that all repeated brackets

$$
[\xi_n, \ldots, [\xi_2, \xi_1]] \cdots]
$$

of length n are identically zero. In terms of the adjoint representation, *Engel's Theorem* says that a Lie algebra is nilpotent if for each X, there exists an exponent n_X such that

$$
(\operatorname{ad}X)^{n_X}(Y) = 0
$$

for all Y in the algebra. The HW algebra is an example.

1. Let $Z = d/dx$, $Y = x^2/2$, $X = x$, $I = 1$. Show that, with commutator as bracket, this is the basis for a nilpotent Lie algebra generated by $\{\,Z, Y\,\}$.

2. Find the group law for the group elements of the form

$$
g(a, b, c, d) = e^{aI}\, e^{bX}\, e^{cY}\, e^{dZ}
$$

3. Try similar studies for Lie algebras generated by $Z = d/dx$ and $Y = x^3/3$, for $Y = x^4/4$, etc.

4. For Problem 2 find the splitting formula using the methods of §IV.

Chapter 2   HYPERGEOMETRIC FUNCTIONS

In this brief chapter, as well as familiarizing ourselves with the notation and the functions, we will see how to use the HW and sl(2) algebras to derive some identities for hypergeometric functions. The general idea is to see some implications of the commutation rules and associative laws for the algebras in particular representations.

I. Notations

The general hypergeometric functions are defined by the series':

$$_pF_q \left(\begin{matrix} a_1, a_2, \ldots, a_p \\ b_1, b_2, \ldots, b_q \end{matrix} \,\middle|\, x \right) = \sum_{n=0}^{\infty} \frac{(a_1)_n (a_2)_n \cdots (a_p)_n}{(b_1)_n (b_2)_n \cdots (b_q)_n} \frac{x^n}{n!}$$

for integer $p, q \geq 0$. They provide a convenient method of standardizing the presentation of special functions. It is important to note that if, say, a_1 is a negative integer, then the series reduces to a finite sum, a polynomial in the remaining a variables and in x. It is also permissible for b_1, say, to be a negative integer as well as a_1, which case will arise in our study, as long as $b_1 \leq a_1 < 0$, since the series will terminate before zero denominators occur.

II. Generating function for $_2F_1$

We start with a formula that will be useful in our work.

2.1 Proposition.   *A generating function for $_2F_1$ polynomials is*

$$(1+x)^m (1+ax)^n = \sum \binom{m+n}{r} x^r \, _2F_1 \left(\begin{matrix} -n, -r \\ -m-n \end{matrix} \,\middle|\, 1-a \right)$$

Proof: Write $1 + ax = 1 + x + (a-1)x$. Then the left side gives:

$$(1+x)^{m+n} (1 + (a-1)x/(1+x))^n = \sum_{k=0}^{n} \binom{n}{k} ((a-1)x)^k (1+x)^{m+n-k}$$

Expand $(1+x)^{m+n-k}$ in series, using our summation convention:

$$\sum_{k=0}^{n} \binom{n}{k} ((a-1)x)^k (-m-n+k)_\alpha \frac{(-x)^\alpha}{\alpha!}$$

Now set $k + \alpha = r$ and combine terms to get the result. It is convenient to replace α by $r - k$ and to use the relation of the form, see Intro., eq. (3.2):

$$(A + k)_{B-k} = (A)_B / (A)_k$$

to reduce the coefficients to standard hypergeometric forms.   ■

Remark. Comparing with direct expansion of the generating function by the binomial theorem yields:

$$\sum \binom{m}{k-r}\binom{n}{r} a^r = \binom{m+n}{k} {}_2F_1\left(\begin{matrix} -k, -n \\ -m-n \end{matrix}\middle| 1-a\right) \tag{2.1}$$

This is useful for writing expressions in terms of the ${}_2F_1$. Cf. reference list of polynomials in the Introduction, §V.

III.   General formulation of CVPS and transformation formulas

The term CVPS is a concatenation of the names Chu-Vandermonde and Pfaff-Saalschütz whose names are customarily attached to the identities we will discuss here: the Chu-Vandermonde formula comes via the HW algebra, the Pfaff-Saalschütz formula, via sl(2). First we present the basic theorem. (Note: here below a and b denote nonnegative integers.)

3.1 Theorem.   *Let L and R satisfy a commutation formula of the type:*

$$L^n R^m = \sum R^{m-j} L^{n-j} S(m, n, j; \rho)$$

where ρ shifts by steps of size ε:

$$\rho R = R(\rho + \varepsilon), \qquad L\rho = (\rho + \varepsilon)L$$

Then the coefficients $S(-\;; \rho)$ satisfy, on the representation space with $\rho\Omega = c\Omega$,

1. *The CVPS identity*

$$S(m+r, n, k; c) = \sum_j S(m, n, j; c + r\varepsilon) S(r, n-j, k-j; c)$$

2. *The transformation formula*

$$\sum_j S(b, n+a-j, N-j; c)\, S(m, n, j; c + (b-a)\varepsilon)$$

$$= \sum_j S(m+b-j, n, N-j; c + (j-a)\varepsilon)\, S(b, a, j; c)$$

Proof: First note that the CVPS identity follows from the transformation formula by setting $a = 0$, $b = r$, $N = k$, since the sum on the right side reduces to one term, with $j = 0$. For the transformation formula, consider $(L^n R^m L^a) R^b = L^n (R^m L^a R^b)$. On the left side, first commute L^n past R^m, then pull the ρ terms through L^a, R^b. Then commute the L terms past R^b. On the right side commute first L^a past R^b. Comparing coefficients yields the formula. ∎

Remark. Note that we can use any real number for r, as long as we have a representation where the operators can act on a vector of the form $R^r \Omega$, which is certainly the case for the HW and sl(2) algebras. E.g., for the HW algebra, $R^r \Omega$ is effectively the function x^r.

IV.   Formulas related to HW algebra

For the HW algebra, Chapter 1, Prop. 2.2.2, gives the form of the coefficients

$$S(m, n, j; 1) = \binom{m, n}{j} \tag{4.1}$$

and the operator $\rho = [L, R] = 1$ commutes with $L = D$ and $R = x$, i.e. the shift $\varepsilon = 0$.

4.1   CV FORMULA

Replacing $S(m, n, j; 1)$ as noted above, eq. (4.1), and setting $\varepsilon = 0$, we have

4.1.1 Proposition.   *The Chu-Vandermonde identity is given by:*

$$\binom{m + r, n}{k} = \sum_j \binom{m, n}{j} \binom{r, n - j}{k - j}$$

or equivalently, by the binomial coefficient identity:

$$\binom{m + r}{k} = \sum_j \binom{m}{j} \binom{r}{k - j} \tag{4.1.1}$$

Note that here m and r can be any real numbers.

Remark. Alternatively, write the binomial terms of eq. (4.1.1) in the form

$$\binom{m}{j} \binom{r}{k - j} = \frac{r!}{(r - k)! \, k!} \frac{(-m)_j (-k)_j}{(r - k + 1)_j \, j!}$$

and divide out by $\binom{r}{k}$ to get

$$_2F_1 \left(\begin{matrix} -m, -k \\ r - k + 1 \end{matrix} \,\middle|\, 1 \right) = \frac{(-m - r)_k}{(-r)_k} \tag{4.1.2}$$

cf. the PS formula of §5.1. This is a special case of the Gauss identity:

$$_2F_1 \left(\begin{matrix} a, b \\ c \end{matrix} \,\middle|\, 1 \right) = \frac{\Gamma(c) \Gamma(c - a - b)}{\Gamma(c - a) \Gamma(c - b)}$$

which holds even for nonterminating series as long as the real parts of the arguments of the Γ–functions are all positive.

4.2 HW TRANSFORMATION FORMULA

Similarly, substituting eq. (4.1) in Theorem 3.1, 2., yields:

4.2.1 Proposition.   *The HW transformation formula is given by:*

$$\sum_j \binom{b, n + a - j}{N - j} \binom{m, n}{j} = \sum_j \binom{m + b - j, n}{N - j} \binom{b, a}{j}$$

or, equivalently, as the $_3F_2$ transformation:

$$_3F_2 \left(\begin{matrix} -N, \, -m, \, -n \\ b - N + 1, \, -n - a \end{matrix} \, \middle| \, 1 \right) =$$

$$\frac{(n - N + 1)_N (-m - b)_N}{(b - N + 1)_N (-n - a)_N} \, _3F_2 \left(\begin{matrix} -N, \, -a, \, -b \\ n - N + 1, \, -m - b \end{matrix} \, \middle| \, 1 \right)$$

V.   Formulas related to sl(2) algebra

For the sl(2) algebra, we use, Chapter 1, Prop. 3.3.3:

$$S(m, n, j; \rho) = \binom{m, n}{j} (\rho + m - n)_j \tag{5.1}$$

Recall Ch. 1, eq. (3.1.1), that ρ acts like twice the number operator (considering commutation rules with Δ and R). That is, for sl(2), $\varepsilon = 2$.

5.1 PS FORMULA

Applying Theorem 3.1, we have:

5.1.1 Proposition.   *The Pfaff-Saalschütz identity is given by*

$$_3F_2 \left(\begin{matrix} -k, \, -m, \, c + 2r + m - n \\ r - k + 1, \, c + r - n \end{matrix} \, \middle| \, 1 \right) = \frac{(-m - r)_k (c + m - n + r)_k}{(-r)_k (c + r - n)_k}$$

Proof: Substitute as noted above for S, eq. (5.1), and put $\varepsilon = 2$. The relations, Intro., eq. (3.2):

$$(A + k)_{B-k} = (A)_B / (A)_k$$

are useful in putting into standard hypergeometric form.   ∎

5.2 sl(2) TRANSFORMATION FORMULA

Similarly, one finds

5.2.1 Proposition.   *The sl(2) transformation formula is given by*

$$
{}_4F_3 \left(\begin{array}{c} -N, \ -m, \ -n, \ c + 2(b-a) + m - n \\ b - N + 1, \ -n - a, \ c + b - n - a \end{array} \middle| 1 \right) =
$$

$$
\frac{(n - N + 1)_N (-m - b)_N (c + b - 2a + m - n)_N}{(b - N + 1)_N (-n - a)_N (c + b - a - n)_N} \times
$$

$$
{}_4F_3 \left(\begin{array}{c} -N, \ -a, \ -b, \ c + b - a \\ n - N + 1, \ -m - b, \ c + b - 2a + m - n \end{array} \middle| 1 \right)
$$

Remark. These results are presented to indicate some of the significance of the Leibniz rules as well as to help explain the significance of these particular identities for hypergeometric functions. They provide some identities that will be of some use in the sequel as well. The study of identities satisfied by hypergeometric functions and their properties in general forms an entire area of study in itself. See Bailey[2], Slater[44], and for a recent treatment of 'basic hypergeometric functions' that is a good indication of the scope of the field in general, see Gasper&Rahman[22].

VI.   Exercises and examples

6.1 EXERCISES

1. Verify equation (2.1).

2. Some properties of hypergeometric functions

   a. Write e^x as a ${}_pF_q$ function.

   b. Show that

   $$
   \cosh x = {}_0F_1 \left(\begin{array}{c} - \\ 1/2 \end{array} \middle| \frac{x^2}{4} \right)
   $$

   c. Show that

   $$
   \frac{\partial}{\partial x} \, {}_0F_1 \left(\begin{array}{c} - \\ c \end{array} \middle| x \right) = c^{-1} \, {}_0F_1 \left(\begin{array}{c} - \\ c + 1 \end{array} \middle| x \right)
   $$

   d. Find formulas similar to that in c. for ${}_1F_1$ and ${}_2F_1$ functions. Find a formula for the general ${}_pF_q$ function.

   e. Use the results of parts b. and c. to give a representation of the function $\sinh x$ in terms of hypergeometric functions.

3. Fill in the proof of Theorem 3.1.

4. Find a proof of equation (4.1.1) by generating functions. Give a combinatorial proof as well.

5. Verify Proposition 4.2.1.

6. Fill in the details of Proposition 5.1.1.

7. In Proposition 5.1.1, what happens if $m = -r$?

8. What happens if $c \to \infty$ in Proposition 5.1.1?

9. Fill in the details for Proposition 5.2.1.

10. Study the limit $c \to \infty$ in Proposition 5.2.1.

6.2 ALGEBRA OF FACTORIAL POWERS

The factorial powers $x^{(n)} = x(x-1)\cdots(x-n+1)$ form a commutative, associative algebra. The 'closed form' summation formulas for hypergeometric functions usually involve products of gamma functions, which extend this algebra to noninteger values of n.

1. Derive the formulas:

$$x^{(m)}x^{(n)} = \sum_k \binom{m}{k}\binom{n}{k} k! \, x^{(m+n-k)}$$

and

$$\binom{x}{m}\binom{x}{n} = \sum_k \binom{m+n-k}{k,m-k,n-k}\binom{x}{m+n-k}$$

2. Write out the product $x^{(l)}x^{(m)}x^{(n)}$ two ways according to associativity. Write in terms of $_3F_2$ functions and deduce relations corresponding to permutation symmetry of the parameters l, m, n.

(See Rademacher[39] for more on the algebra generated by binomial coefficients.)

Chapter 3 PROBABILITY AND FOCK SPACES

The concept of probability involves two basic notions: that of *positivity* and the paramount concept of *average* or *mathematical expectation*. Typically the quantities one wants to average form an algebra. We want to be able to add, multiply, and scale random variables. Constants should be included as well, as deterministic quantities. For example, one wants to be able to calculate means and variances of random variables, and correlations between random variables. In algebraic terms, from the point of view of functional analysis, these become the notions: positive functionals, squared norms, and inner products.

We begin with the rudiments of how these notions may be defined on operator algebras. The discussion is put in the finite-dimensional setting, but extends practically verbatim to appropriate classes of operators on Hilbert space. In the situations of primary interest, in Chapter 5, we will have operators defined on a polynomial basis for a Hilbert space. We will do calculations explicitly, thus providing interesting illustrations of the theory.

Remark. In this chapter, we present the discussion for complex vector spaces, as the restriction to the real case will be clear. We use * to denote the Hermitian transpose of a complex matrix, i.e. conjugate transpose. In agreement, on real matrices, the * means transpose. Bar will denote complex conjugate.

I. Trace functionals: probability and operators

The notion of 'random quantities' here is an operator algebra on which one can define an inner product and for which one has a notion of positivity. The principal distinguishing feature of the formulation in terms of operator algebras is that the quantities do not necessarily commute. The 'classical theory' of probability corresponds to choosing a commutative algebra of, say, bounded measurable functions on the real line (bounded random variables). We will consider an algebra of operators, e.g. an algebra of matrices, on a real or complex vector space.

1.1 INNER PRODUCTS

First we recall the basic definitions.

1.1.1 Definition. On a complex vector space $\mathcal{V}$ an *inner product* is a mapping $\langle \cdot, \cdot \rangle : \mathcal{V} \times \mathcal{V} \to \mathbf{C}$ such that, for $u, v, w \in \mathcal{V}$, $\alpha, \beta \in \mathbf{C}$:

1. $\langle \alpha u + v, \beta w \rangle = \alpha \bar{\beta} \langle u, w \rangle + \bar{\beta} \langle v, w \rangle$

2. $\langle u, v \rangle = \overline{\langle v, u \rangle}$

3. $\langle u, u \rangle > 0$, unless $u = 0$

As usual we denote $\langle v, v \rangle$ by $\|v\|^2$, the *squared norm* . And we recall *Schwartz'*
inequality

$$|\langle u, v \rangle| \leq \|u\| \, \|v\|$$

To define an inner product on operators, we use the *trace*. It is easy to see
that setting

$$\langle X, Y \rangle = \mathrm{tr}\,(XY^*)$$

gives a positive-definite inner product for operators on a finite-dimensional vector
space (and, suitably interpreted and restricted, for infinite-dimensional spaces as
well).

Remark. It is also interesting to consider the bilinear form $\mathrm{tr}\,(XY)$. See §4.2
below. In the theory of Lie algebras, this leads to the *Killing form*.

Properties of the trace yield properties of this inner product. We note:

1.1.2 Proposition.   *The inner product is consistent with the adjoint operation:*

$$\langle AX, Y \rangle = \langle X, A^*Y \rangle, \qquad \langle X, YA \rangle = \langle XA^*, Y \rangle$$

One way to see how the inner product works is to observe that $\langle X, Y \rangle$ is the
ordinary inner product of X, Y considering the matrices as n^2-dimensional vectors:

$$\langle X, Y \rangle = \sum_{j,k} x_{jk} \bar{y}_{jk}$$

1.1.1 Pauli matrices

For 2×2 complex matrices one has a special basis, the *Pauli matrices* . The
Pauli matrices, σ_j, are used in the quantum theory of spin. They are 2×2 matrices
defined by:

$$\sigma_0 = I = \begin{pmatrix} 1 & 0 \\ 0 & 1 \end{pmatrix}, \quad \sigma_1 = \begin{pmatrix} 0 & 1 \\ 1 & 0 \end{pmatrix}, \quad \sigma_2 = \begin{pmatrix} 0 & -i \\ i & 0 \end{pmatrix}, \quad \sigma_3 = \begin{pmatrix} 1 & 0 \\ 0 & -1 \end{pmatrix}$$

The matrices $\{\sigma_1, \sigma_2, \sigma_3\}$ give a basis of Hermitian matrices for sl(2). They are
an orthogonal system with respect to the inner product defined above. They have
many interesting properties. (Recall the Kronecker and Levi-Civita symbols, see
§II of the Introduction.)

1.1.1.1 Proposition.   *The matrices σ_j ($j = 1, 2, 3$) satisfy:*

1. *Commutation relations*

$$[\sigma_j, \sigma_k] = \varepsilon_{jkl}\, 2i\sigma_l$$

2. *Anticommutation relations*

$$\sigma_j \sigma_k + \sigma_k \sigma_j = \delta_{jk}\, 2\sigma_0$$

3. *Orthogonality*

$$\langle \sigma_j, \sigma_k \rangle = 2\delta_{jk}$$

Proof: The proofs are straightforward. Note that we can compute the inner products from the anticommutation relations. Since these are Hermitian matrices, we have only to take the trace:

$$2 \langle \sigma_j, \sigma_k \rangle = \operatorname{tr} (\sigma_j \sigma_k + \sigma_k \sigma_j) = 2\delta_{jk} \operatorname{tr} I$$

and divide out the factor of 2. ∎

We note the relation with the standard basis of Ch.1:

$$R = \tfrac{1}{2}(\sigma_1 + i\sigma_2), \qquad \Delta = -\tfrac{1}{2}(\sigma_1 - i\sigma_2), \qquad \rho = \sigma_3$$

1.2 EXPECTATION FUNCTIONALS

An expectation functional is the mathematical formulation of the notion of average. In the following definition, the elements f and g belong to an algebra with identity over **R** or **C**. The scalar c denotes equally the algebra element c times the identity.

1.2.1 Definition. A functional defined on an algebra is an *expectation* (denoted by angle brackets $\langle \rangle$) if it satisfies:

1. It is linear: $\langle f + cg \rangle = \langle f \rangle + c \langle g \rangle$, with c any scalar.
2. It preserves constants: $\langle c \rangle = c$ for any constant c.
3. It is positivity-preserving: If $f \geq 0$, then $\langle f \rangle \geq 0$.

For the algebra of functions on **R**, such a functional is given by integration with respect to a probability measure — a positive measure of total mass one. (See the next section.)

We need to specify the notion of positivity in the operator context. Typically, a self-adjoint matrix **p** acting on an inner product space $\mathcal{V}$, with inner product $(\cdot, \cdot)$, is said to be positive if $(\mathbf{p}v, v) \geq 0$, for all $v \in \mathcal{V}$. Note that with the $\geq$ sign, this is often called positive-semidefiniteness. We use the terminology *positive* to mean positive-semidefinite, non-zero, and then *strictly positive* means: $(\mathbf{p}v, v) > 0$ unless $v = 0$.

Remark. The main fact we need about positive matrices is:

a necessary and sufficient condition for a matrix X to be positive is that there exist a nonzero matrix a such that $X = a^* a$

In terms intrinsic to the operator algebra, one has

1.2.2 Definition. A non-zero element **p** of an operator algebra is *positive* if $\mathbf{p} = \mathbf{p}^*$ and the spectrum of **p** is nonnegative. I.e., it is nonnegative, self-adjoint.

Consistent with standard terminology, we make

1.2.3 Definition.   A positive operator **p** of trace one is a *density matrix* .

These are the operator correlates of probability measures. The terminology is consistent with the notion of a probability density on **R** (see below).

1.2.4 Proposition.   *For any density matrix* **p**,

$$\langle X \rangle_{\mathbf{p}} = \mathrm{tr}\,(X\mathbf{p})$$

defines an expectation functional on the algebra of matrices.

Proof: The first two properties of Def. 1.2.1 are immediate. For positivity we have to check that if X is itself positive, then $\langle X \rangle_{\mathbf{p}} \geq 0$. Write $X = a^*a$ and $\mathbf{p} = rr^*$. Then, in terms of the inner product, using Prop. 1.1.2 we have:

$$\langle X \rangle_{\mathbf{p}} = \langle a^*a, rr^* \rangle = \langle a, arr^* \rangle = \langle ar, ar \rangle \geq 0$$

∎

1.3   PROBABILITY MEASURES

Throughout, we will be concerned with probability measures only on (subsets of) **R**. A probability measure, conventionally denoted $p(dx)$, is determined by a non-decreasing function, $F(x)$, the *distribution function* such that $F(x) \to 0$ as $x \to -\infty$ and $F(x) \to 1$ as $x \to +\infty$. (One takes F to be right-continuous by convention.) The corresponding measure $p(dx)$ is defined so that

$$F(b) - F(a) = \int_{(a,b]} p(dx) = p((a,b])$$

for any interval $(a, b] \subset \mathbf{R}$.

Remark. It is important to note that a probability measure $p(dx)$ is determined by the corresponding expectation functional

$$\langle f \rangle = \int_{-\infty}^{\infty} f(x)\, p(dx)$$

on the algebra of *bounded continuous functions* on **R**.

We are particularly interested in measures given either by density functions or point masses. Precisely

1.3.1 Definition. A *density function* $p(x)$ is a nonnegative function on **R**, $p(x) \geq 0, \forall x$, such that

$$\int_{-\infty}^{\infty} p(x)\, dx = 1$$

In this case, we have $p(dx) = p(x)\, dx$. The corresponding expectation is defined by integration:

$$\langle f \rangle = \int f(x) p(x)\, dx$$

for bounded continuous functions f. For discrete probabilities, for example, for random variables taking only integer values, we need point masses. First we have

1.3.2 Definition. The *delta function* at the point a is the probability measure $\delta_a(dx)$ defined by

$$\int_{-\infty}^{\infty} f(x)\, \delta_a(dx) = f(a)$$

for any continuous function f on **R**.

The terminology *point masses* refers to weighted delta functions. A *discrete probability measure* is given by point masses with total weight equal to one.

Remark. The terms probability distribution, distribution, and probability measure are generally used interchangeably.

Some examples of discrete distributions:

1. A *Bernoulli distribution* is the sum of two point masses. Take $0 < p, q < 1$, two numbers such that $p + q = 1$. Pick two points $a, b \in \mathbf{R}$. Then

$$p(dx) = p\, \delta_a(dx) + q\, \delta_b(dx)$$

   is a probability measure — the Bernoulli distribution corresponding to choosing points a and b with probabilities p and q respectively. The corresponding expectation on functions f is given by: $\langle f \rangle = p\, f(a) + q\, f(b)$.

2. A *Poisson distribution* is a distribution on the nonnegative integers determined by a parameter $\lambda > 0$:

$$p_\lambda(dx) = e^{-\lambda} \sum_{n=0}^{\infty} \frac{\lambda^n}{n!}\, \delta_n(dx)$$

   with the corresponding expectation:

$$\langle f \rangle = e^{-\lambda} \sum_{n=0}^{\infty} \frac{\lambda^n}{n!}\, f(n)$$

3. In general, a discrete distribution has the form

$$p(dx) = \sum_{n=0}^{\infty} p_n \, \delta_{a_n}(dx)$$

where $p_n \geq 0$, $\forall n$, $\sum p_n = 1$, and $\{a_n\}$ is a given sequence of real numbers. The corresponding expectation is

$$\langle f \rangle = \sum_{n=0}^{\infty} p_n \, f(a_n)$$

Remark. Occasionally we use the notation $\mu(dx)$ to denote a positive, finite measure with total mass not necessarily equal to 1. One could normalize such a $\mu(dx)$ by dividing by $\mu((-\infty, +\infty))$ to get a probability measure.

1.4 BOCHNER'S THEOREM

To see the correspondence between the operator-theoretic and measure-theoretic viewpoint, we employ *Bochner's theorem* . First, we have

1.4.1 Definition. A (complex-valued) function on **R** is *positive-definite* if, for all $n \geq 1$, for all finite subsets $\{t_1, \ldots, t_n\}$ of **R** and complex vectors $(\xi_1, \ldots, \xi_n)$, one has

$$\sum_{1 \leq j,k \leq n} f(t_j - t_k)\xi_j \bar{\xi}_k \geq 0$$

I.e., the matrix $P_{jk} = f(t_j - t_k)$ is positive for every $n \geq 1$ and corresponding t's. Then *Bochner's theorem* states:

if f is positive-definite and continuous at 0, then it is the Fourier transform of a positive measure

i.e., there is a positive (finite) measure $\mu(dx)$ such that

$$f(t) = \int_{-\infty}^{\infty} e^{itx} \, \mu(dx)$$

In particular, if $f(0) = 1$, μ must be a probability measure, which we then denote by $p(dx)$.

We can now see the connection between operators and measures. The theorem, stated for the finite-dimensional case, holds for self-adjoint operators on Hilbert space as well.

1.4.2 Theorem. *As a functional on self-adjoint matrices, the expectation $\langle \cdot \rangle_{\mathbf{P}}$ can be expressed in terms of probability distributions on **R**.*

Proof: Let X be self-adjoint and let $\mathbf{p}$ be a density matrix. Define the function

$$f(t) = \langle e^{itX} \rangle_{\mathbf{p}}$$

Since we are in the finite-dimensional case, f is continuous. And $f(0) = \langle I \rangle_{\mathbf{p}} = 1$. We check that $f(t)$ is positive definite:

$$\sum_{j,k} f(t_j - t_k)\xi_j\bar{\xi}_k = \sum_{j,k} \langle e^{i(t_j - t_k)X} \rangle_{\mathbf{p}} \xi_j\bar{\xi}_k = \langle \Big(\sum_j e^{it_j X} \xi_j\Big)\Big(\sum_j e^{it_j X} \xi_j\Big)^* \rangle_{\mathbf{p}} \geq 0$$

as required. By Bochner's theorem, there is a corresponding probability measure $p(dx)$, depending on $\mathbf{p}$ and X, such that

$$f(t) = \langle e^{itX} \rangle_{\mathbf{p}} = \int_{-\infty}^{\infty} e^{itx} p(dx)$$

∎

For fixed $\mathbf{p}$, associated to each X is a probability measure, the *spectral measure*.

In the finite-dimensional case, with $\mathbf{p}$ the identity matrix normalized to trace one, corresponding to X is a discrete measure with point masses at its spectrum — the eigenvalues of X. Generally, intervals with positive probability with respect to the measure $p(dx)$ contain points of the spectrum of X. In the infinite-dimensional case, $p(dx)$ can have a density $p(x)$, corresponding to operators X with continuous spectrum.

1.5 GNS FORM

As we saw in the proof of Prop. 1.2.4, via the factorization of the density matrix $\mathbf{p} = rr^*$, we can express the expectation $\langle \cdot \rangle_{\mathbf{p}}$ in terms of the inner product:

$$\langle X \rangle_{\mathbf{p}} = \langle X, rr^* \rangle = \langle Xr, r \rangle$$

Conversely, given a unit operator r, $\langle r, r \rangle = 1$, we can form $\mathbf{p} = rr^*$, which will be a density operator, hence determining an expectation. In fact, we have for any space on which the algebra acts:

1.5.1 Proposition.   *To each unit vector r in a representation space corresponds an expectation functional:*

$$\langle X \rangle_r = \langle Xr, r \rangle$$

Proof: Positivity here may be directly checked, cf. Prop. 1.2.4, writing $X = a^*a$ for positive X so that:

$$\langle X \rangle_r = \langle a^*a \rangle_r = \langle ar, ar \rangle \geq 0$$

Note that a unit vector is required to have for the identity $\langle I \rangle_r = 1$.   ∎

The GNS — Gel'fand-Naimark-Segal — theorem is the converse result in the infinite-dimensional setting. Briefly,

   the form $\langle Xr, r \rangle$ is generic for expectations.

II. Fock spaces

Recall from Chapter 1, §§2.2.2, 3.3.1, that to get a 'calculus' for the Lie algebra we construct a vector space by applying the algebra to a vacuum vector Ω yielding a basis ψ_n. The corresponding *Fock space* is a Hilbert space constructed so that the ψ_n form an orthogonal basis.

Fock space (for bosons) as usually defined is constructed from an *infinite* set of boson pairs (Ch. 1, Def. 2.2.2.1): $\{a_k, a_k^\dagger\}$, for integer k, $-\infty < k < \infty$. The basis vectors are built by applying the *creation operators* $a_k^\dagger$ to a vacuum vector Ω. The *annihilation operators* a_k all satisfy $a_k\Omega = 0$.

We are interested in working with a *finite* set of generators. To present the idea behind our work, we outline the general features we are looking for. For us, a Fock space Φ comes from the following type of construction:

1. Start with a finitely generated associative algebra $\mathcal{A}$ and a vacuum vector Ω

2. As a vector space, Φ consists of elements of the form $\xi\Omega$, $\xi \in \mathcal{A}$

3. Denote a set of generators (as an associative algebra) of $\mathcal{A}$ by Ξ. The Hilbert space structure of Φ is such that Ξ can be written as the union of three subsets

$$\Xi = \Xi_+ \cup \Xi_0 \cup \Xi_-$$

    with these properties:
    a. Ξ_- consists of the adjoints of Ξ_+
    b. Elements of Ξ_0 act as multiplication by scalars on Ω
    c. Elements of Ξ_- annihilate Ω
    d. For any $\xi \in \Xi_+$, the commutator with its adjoint $\xi^* \in \Xi_-$ gives an element $[\xi, \xi^*] \in \Xi_0$

4. The operators Ξ_+ are the *raising operators* and the Ξ_- are the *lowering operators*

Remark. Every element of the form $[\xi, \xi^*]$ is self-adjoint. However, we do not require every element of Ξ_0 to be self-adjoint. It is natural, however, to require that the *set* Ξ_0 be self-adjoint, i.e., that $\xi \in \Xi_0 \Rightarrow \xi^* \in \Xi_0$. One can thus extend to the case where Ω is replaced by a finite-dimensional representation space of Ξ_0.

We will work with the HW and sl(2) algebras of Chapter 1. The corresponding spaces Φ will be said to be of HW-type (§2.3) or of sl(2)-type (§2.4), respectively. We will find these to possess interesting features sufficient for the present study.

2.1 RAISING AND LOWERING OPERATORS. V OPERATOR

Here is our formulation. We have one raising operator R, with corresponding lowering operator $L = R^*$. The basis for Φ is given by $\psi_n = R^n \Omega$, $n \geq 0$, such that

$$\langle \psi_n, \psi_m \rangle = \delta_{nm} \gamma_n$$

with $\gamma_n = \|\psi_n\|^2$ denoting the squared norms. The action of L is given by

$$L\psi_n = b_n \psi_{n-1}, \qquad n \geq 1$$

with $L\Omega = 0$, where the sequence $\{\, b_n \,\}$ is determined by the Lie algebra structure (see Ch. 1, eq. (2.2.2.1), Props. 2.2.2.3, 3.3.1.1, and 3.5.3). We define the *velocity operator* V by the action

$$V\psi_n = n\psi_{n-1}, \qquad n \geq 1$$

with $V\Omega = 0$. Note that V, R are a boson pair (Ch. 1, Def. 2:2.2.1).

In this chapter we will work with a realization of Φ as a space of functions. Make the isometric vector space correspondence:

$$x^n \quad \leftrightarrow \quad \psi_n$$

This is extended by linearity to polynomials. In this realization by functions, the Hilbert space Φ is determined by the requirement that the functions x^n, $n \geq 0$, are an orthogonal basis with squared norms $\gamma_n = \|x^n\|^2$.

Remark. It is essential to warn the reader in advance that this is *not* the same 'x' as in Chapters 5, 6, and 7. There, a different realization by functions is given, while the operators R, L, and V retain their significance throughout.

Finally, we make a notational convention:

Remark. We will consistently use 'x' as the variable for functions in Φ when using the notation $\langle \cdot, \cdot \rangle$. For example, in the expression $\langle f(x,y), g(x,y) \rangle$ it is understood that for fixed y these are elements in Φ, and the variable x is 'integrated out', resulting in a function of y.

2.2 NORMS AND REPRODUCING KERNELS

We discuss in this section two important features.

2.2.1 *Norms*

From the construction of Φ, we find the significance of the sequence $\{ b_n \}$ for the Hilbert space structure.

2.2.1.1 Proposition. *For $n \geq 1$, the squared norms are given by*

$$\gamma_n = \|\psi_n\|^2 = b_1 b_2 \cdots b_n \|\Omega\|^2$$

Proof: Since L and R are adjoint, we have, for $n \geq 1$:

$$\gamma_n = \langle \psi_n, \psi_n \rangle = \langle R\psi_{n-1}, \psi_n \rangle$$
$$= \langle \psi_{n-1}, L\psi_n \rangle = b_n \gamma_{n-1}$$

and the result follows by induction. $\blacksquare$

We state

2.2.1.2 Corollary. *Up to the factor $\|\Omega\|^2$, the sequence $\{ b_n \}$ determines the sequence $\{ \gamma_n \}$ of squared norms and vice versa.*

Notice that this correspondence relies on the Fock-space structure.

2.2.2 *Reproducing kernels*

In this subsection, we consider any Hilbert space of functions, $\{ f(x) \}$, generally complex-valued, with orthogonal basis $\{ \psi_n(x) \}$, satisfying $\gamma_n = \|\psi_n\|^2$, $n \geq 0$. The functions $f(x)$ are determined by the condition that they admit an expansion of the form $\sum a_n \psi_n$ with

$$\|f\|^2 = \sum |a_n|^2 \gamma_n < \infty$$

This expansion converges in the Hilbert space norm:

$$f_N(x) = \sum_{n=0}^{N} a_n \psi_n(x)$$

satisfies

$$\lim_{N \to \infty} \|f_N - f\|^2 = 0$$

Taking inner products we see that the coefficients a_n are given by

$$a_n = \langle f, \psi_n \rangle / \gamma_n \tag{2.2.2.1}$$

Now, introduce a (dual) copy of the space with orthogonal basis $\{ \psi_n(y) \}$. Consider the formal sum

$$k(x, y) = \sum_{n=0}^{\infty} \frac{\psi_n(x)\overline{\psi_n(y)}}{\gamma_n} \tag{2.2.2.2}$$

This is a *weak reproducing kernel* . It 'reproduces' functions f in the space in the following sense. For fixed y, let

$$k_N(x,y) = \sum_{n=0}^{N} \frac{\psi_n(x)\overline{\psi_n(y)}}{\gamma_n}$$

Then

$$f_N(y) = \langle f(x), k_N(x,y)\rangle = \sum_{n=0}^{N} \psi_n(y)\,\langle f, \psi_n\rangle/\gamma_n$$

as in eq. (2.2.2.1), converges to $f(y)$ in norm, as noted above.

In the case where the sum (2.2.2.2) converges to an element in the space, we have a *reproducing kernel* denoted here by $K(x,y)$. The Hilbert space is then called a *reproducing kernel Hilbert space* . We require that

$$\langle K(x,y), K(x,y)\rangle = \sum_{n=0}^{\infty} \frac{|\psi_n(y)|^2}{\gamma_n} < \infty$$

for all y in the domain of the functions ψ_n. Thus,

$$f(y) = \langle f(x), K(x,y)\rangle$$

has direct meaning in the sense of the Hilbert space inner product. This leads immediately to bounds for the functions $f(x)$.

2.2.2.1 Proposition.   *The pointwise bound*

$$|f(x)| \leq \sqrt{K(x,x)}\,\|f\|$$

holds for all f in the reproducing kernel Hilbert space.

Proof: $K_y(x) = K(x,y)$ is an element of the Hilbert space for each fixed y. Then, by the reproducing property:

$$K(y,y) = K_y(y) = \langle K_y(x), K(x,y)\rangle = \|K_y\|^2$$

Thus, applying Schwartz' inequality to

$$f(y) = \langle f(x), K(x,y)\rangle = \langle f, K_y\rangle$$

yields

$$|f(y)| \leq \|K_y\|\,\|f\|$$

and, with the replacement of y by x, the result follows.   ■

Now we are ready to look at the Fock spaces realized by the HW and sl(2) algebras in turn. Recalling Corollary 2.2.1.2, we remark that the spaces may be identified by their norms. These will serve as convenient indicators when we study the Bernoulli systems in Chapter 5.

2.3 HW–TYPE SPACES

Let V, R be a boson pair. With $L = hV$, $h > 0$, the corresponding Fock space Φ_h is said to be of *HW-type* . Recall Ch. 1, Prop. 2.2.2.3, that the action of L on the basis $\psi_n = R^n \Omega$ is given by

$$L\psi_n = h\, n\psi_{n-1}$$

Thus,

2.3.1 Proposition.   *For the HW-type space, we have the coefficients b_n given by $b_n = hn$. The corresponding squared norms are*

$$\gamma_n = h^n n! \, \|\Omega\|^2$$

Proof: This follows directly from Prop. 2.2.1.1.   ∎

Remark. One can construct *orthonormal* states

$$\tilde{\psi}_n = \psi_n / (\sqrt{h^n n!} \, \|\Omega\|)$$

which are often useful, as will be seen in §III.

According to our realization as functions, we take x^n, $n \geq 0$, as an orthogonal basis. With $\|x^n\|^2 = \gamma_n = h^n n!$, normalizing $\|x^0\| = 1$, elements of the Fock space have the form of power series.

2.3.2 Definition.   The Fock space Φ_h denotes functions given by power series

$$f(x) = \sum_{n=0}^{\infty} \frac{a_n x^n}{n!} \tag{2.3.1}$$

with finite norm $\|f\|_h$ given by

$$\|f\|_h^2 = \sum_{n=0}^{\infty} \frac{|a_n|^2 h^n}{n!} < \infty$$

The inner product is thus

$$\langle f, g \rangle = \sum_{n=0}^{\infty} \frac{a_n \bar{b}_n h^n}{n!}$$

for $f = \sum a_n x^n / n!$, $g = \sum b_n x^n / n!$.

Remark. Note that these spaces are nested: $h > h'$ implies $\Phi_h \subset \Phi_{h'}$.

As in Prop. 2.2.2.1, we can derive pointwise bounds for functions in Φ_h.

2.3.3 Proposition.   If $f \in \Phi_h$, then f is an entire function. Furthermore, we have the bound, for all $x \in \mathbf{C}$:

$$|f(x)| \leq e^{|x|^2/2h} \|f\|_h$$

Proof: Denote $|x|$ by r. Apply Schwartz' inequality to the expansion of f, eq. (2.3.1), as follows:

$$\sum_{n=0}^{\infty} \frac{|a_n| r^n}{n!} = \sum_{n=0}^{\infty} \frac{|a_n| h^{n/2} (r/\sqrt{h})^n}{n!} \leq \left(\sum_{n=0}^{\infty} \frac{|a_n|^2 h^n}{n!} \right)^{1/2} \left(\sum_{n=0}^{\infty} \frac{(r^2/h)^n}{n!} \right)^{1/2}$$

and the result follows.   ∎

Now we will see that in fact we do have reproducing kernel Hilbert spaces.

2.3.1 Reproducing kernel

From the theory of §2.2.2, we have

2.3.1.1 Proposition. Φ_h has a reproducing kernel, which is given by the exponential function:

$$e^{x\bar{y}/h}$$

Proof: With $\psi_n(x) = x^n$ and $\gamma_n = h^n n!$, this follows from the expansion (2.2.2.2) as soon as we verify that it gives a *bona fide* element of Φ_h for fixed y. We have

$$\|c^{x\bar{y}/h}\|_h^2 = \sum_{n=0}^{\infty} \frac{h^n n! |y|^{2n}}{h^{2n} n! \, n!} = e^{|y|^2/h} < \infty$$

and the result follows.   ∎

A corollary of this is Prop. 2.3.3, via Prop. 2.2.2.1.

2.3.2 Operator calculus

We can develop an operator calculus for the operator $D = d/dx$, using the spaces Φ_h. Here we go back to the specific realization of V as D, $L = hD$. And further, we will take $h = 1$ and denote the norm in Φ_1 by $\| \cdot \|$.

Our technique is to integrate the reproducing kernel relation with respect to probability measures. Let $p(dx)$ be a probability measure on $\mathbf{R}$. We set, for $x \in \mathbf{C}$,

$$\phi(x) = \int_{-\infty}^{\infty} e^{xy} \, p(dy)$$

This is the *Fourier-Laplace* transform of the measure p. We want this to be an element of Φ_1. Observe that if $x \in \mathbf{R}$, then $\phi(x) > 0$ on its domain of definition.

2.3.2.1 Proposition.   If the Fourier-Laplace transform of p, $\phi(x)$, exists for all $x \in \mathbf{R}$, then $\phi \in \Phi_1$ if and only if

$$\int_{-\infty}^{\infty} \phi(x)\, p(dx) < \infty$$

Proof:

$$\langle \phi, \phi \rangle = \langle \phi(x), \int_{-\infty}^{\infty} e^{xy}\, p(dy) \rangle$$

$$= \int_{-\infty}^{\infty} \langle \phi(x), e^{xy} \rangle\, p(dy)$$

$$= \int_{-\infty}^{\infty} \phi(y)\, p(dy)$$

by the reproducing property of $\exp(xy)$ for real y. Since the integrands are non-negative, the interchange of integration is justified by Fubini's theorem.   ∎

2.3.2.2 Lemma.   For $f \in \Phi_1$,

$$\langle f, 1 \rangle = f(0)$$

Proof: Using the reproducing kernel, evaluate

$$f(y) = \langle f(x), e^{x\bar{y}} \rangle$$

at zero.   ∎

Next we look at translates $f_y(x) = f(x + y)$ of functions in Φ_1 — the action of the translation group on elements of Φ_1. Recall that $\Phi_{2+\varepsilon} \subset \Phi_1$, $\forall \varepsilon > 0$.

2.3.2.3 Lemma.   If, for some $\varepsilon > 0$, $f \in \Phi_{2+\varepsilon}$, then the translates f_y are in Φ_1 for all $y \in \mathbf{R}$.

Proof: Let $f(x) = \sum a_n x^n / n!$. Then $f(x + y) = \sum a_n (x + y)^n / n!$. We have

$$\|(x+y)^n\|^2 = \|\sum_{k=0}^{n} \binom{n}{k} y^{n-k} x^k\|^2$$

$$= \sum_{k=0}^{n} \binom{n}{k}^2 y^{2(n-k)} k!$$

$$= n! \sum_{k=0}^{n} \binom{n}{k} \frac{(y^2)^{n-k}}{(n-k)!} \leq e^{y^2} n!\, 2^n$$

Now, $f_y \in \Phi_1$ if the series of norms converges, i.e., if $\sum |a_n| \, \|(x+y)^n\|/n! < \infty$. We have, via the above estimate and Schwartz' inequality:

$$\sum_{n=0}^{\infty} \frac{|a_n|}{n!} \|(x+y)^n\| \le e^{y^2/2} \sum_{n=0}^{\infty} \frac{|a_n| \, 2^{n/2}}{(n!)^{1/2}}$$

$$= e^{y^2/2} \sum_{n=0}^{\infty} \frac{|a_n| \, (2+\varepsilon)^{n/2}}{(n!)^{1/2}} \left(\frac{2}{2+\varepsilon}\right)^{n/2}$$

$$\le e^{y^2/2} \|f\|_{2+\varepsilon} \times ((2+\varepsilon)/\varepsilon)^{1/2}$$

with $\| \cdot \|_{2+\varepsilon}$ denoting the norm in the space $\Phi_{2+\varepsilon}$. ■

We now study operators $\phi(D)$. Recall that for the measure $p(dx)$, the corresponding expectation is given by $\langle f \rangle = \int f(x) \, p(dx)$, for all f for which the integral converges absolutely.

2.3.2.4 Theorem. *Let p be a probability measure which has Fourier-Laplace transform $\phi \in \Phi_1$. Let $f \in \Phi_1$ be such that $\langle f \rangle$ exists with respect to p. Then*

1. *We can write*

$$\langle f \rangle = \langle f, \phi \rangle = \phi(D) \, f(0)$$

2. *The operator $\phi(D)$ is given by*

$$\phi(D) \, f(x) = \int_{-\infty}^{\infty} f(x+y) \, p(dy)$$

for all x such that the translate $f_x(y) = f(x+y) \in \Phi_1$ as a function of y. In particular, for $f \in \Phi_{2+\varepsilon}$, this holds for all $x \in \mathbf{R}$.

   Proof: Integrate the relation $f(y) = \langle f(x), \exp(x\bar{y}) \rangle$ with respect to $p(dy)$. This yields, by Fubini's theorem,

$$\langle f \rangle = \langle f, \phi \rangle$$

Since on Φ_1, x and D are adjoints, we have

$$\langle f, \phi \rangle = \langle \phi(D) \, f(x), 1 \rangle$$

Applying Lemma 2.3.2.2, statement 1. follows. For 2., apply statement 1. to a translate $f_a(x) = f(a+x)$, a fixed. Then replace a by x in the result. Lemma 2.3.2.3 implies the statement for functions in $\Phi_{2+\varepsilon}$. ■

Remark. The integral in statement 2. is a type of *convolution integral* . We will look at these in more detail in Chapter 4. The operators $\phi(D)$ are referred to as *convolution operators* .

From 1. of Theorem 2.3.2.4, we see

2.3.2.5 Corollary. *For f and g polynomials,*

$$\langle f, g \rangle = f(D)g(0) = \bar{g}(D)f(0)$$

Some further remarks:

1. Theorem 2.3.2.4 suggests the following

2.3.2.6 Definition. The operator $\phi(D)$ is defined via the convolution-type integral

$$\phi(D) f(x) = \int_{-\infty}^{\infty} f(x + y) \, p(dy) \qquad (2.3.2.1)$$

for all x and f for which the integral makes sense.

In particular, for any bounded function f, this is well-defined on all of **R**.

2. The HW formula

$$e^{yD} f(x) = f(x + y)$$

is closely connected with the discussion here. Writing the operator relation

$$\phi(D) = \int_{-\infty}^{\infty} e^{yD} \, p(dy)$$

and applying this to $f(x)$ recovers the basic relation eq. (2.3.2.1).

It is worthwhile to consider this scheme carefully, as it is fundamental to our approach. And we want to see clearly how the properties of the Fock space and the reproducing kernel fit into the picture. So, what are some interpretations of the formula $e^{yD} f(x) = f(x + y)$?

1. Via HW algebra/group laws: the HW commutation rules give

$$e^{yD} f(x) = f(x + y) e^{yD}$$

as operators on polynomials, say. Applying this to the vacuum, here given by the constant function 1, gives the desired relation.

2. For analytic functions, we can expand the exponential

$$e^{yD} f(x) = \sum_{n=0}^{\infty} \frac{y^n D^n}{n!} f(x) = \sum_{n=0}^{\infty} \frac{y^n f^{(n)}(x)}{n!}$$

which equals $f(x + y)$ by Taylor's formula.

3. In the reproducing kernel Hilbert space, we have:

$$f(a + b) = \langle f(x), e^{\overline{x(a+b)}} \rangle$$
$$= \langle f(x), e^{x\bar{a}} e^{x\bar{b}} \rangle$$

and, since x and D are adjoints,

$$= \langle e^{aD} f(x), e^{x\bar{b}} \rangle$$
$$= e^{aD} f(b)$$

Note that this last step entails $e^{aD} f(x) \in \Phi_1$.

2.4 sl(2)–TYPE SPACES

Now we look at the Fock spaces constructed using the sl(2) algebra. We use the standard sl(2) algebra acting on the basis ψ_n with the actions:

$$R\psi_n = \psi_{n+1}, \qquad \rho\psi_n = (c + 2n)\psi_n, \qquad L\psi_n = n(c + n - 1)\psi_{n-1} \qquad (2.4.1)$$

i.e., as in Ch. 1, Prop. 3.5.3, with $\beta = 1$, cf. also, Ch. 1, Props. 3.3.1.1, 3.3.1.2. Thus,

2.4.1 Proposition. *For the sl(2)-type space, we have the coefficients* $b_n = n(c + n - 1)$. *The corresponding squared norms are*

$$\gamma_n = n!\,(c)_n \,\|\Omega\|^2$$

where $(c)_n = \Gamma(c + n)/\Gamma(c)$.

Proof: This follows from Prop. 2.2.1.1.   ∎

The orthonormal states have the form

$$\tilde{\psi}_n = \psi_n/(\sqrt{n!(c)_n}\,\|\Omega\|)$$

Remark. It is important to note that here we can have not only any value of $c > 0$, but negative integer values of c as well. What happens is that the space becomes finite-dimensional — all basis elements ψ_n with $n > -c$ have zero norm, hence are zero in the Hilbert space. In the discussion here we will consider the case $c > 0$, with the modifications for negative c left aside for the present, since in such a case the realization in terms of functions reduces to a space of polynomials of bounded degree. We will study this case explicitly in Chapter 5.

For the realization as functions, we take x^n, $n \geq 0$, as an orthogonal basis such that $\|x^n\|^2 = \gamma_n = n!\,(c)_n$, normalizing $\|x^0\| = 1$. Thus

2.4.2 Definition. For $c > 0$, the Fock space Φ_c denotes functions given by power series

$$f(x) = \sum_{n=0}^{\infty} \frac{a_n x^n}{n!\,(c)_n}$$

with finite norm $\|f\|_c$ given by

$$\|f\|_c^2 = \sum_{n=0}^{\infty} \frac{|a_n|^2}{n!\,(c)_n} < \infty$$

The inner product is thus

$$\langle f, g \rangle = \sum_{n=0}^{\infty} \frac{a_n \bar{b}_n}{n!\,(c)_n}$$

for $f = \sum a_n x^n/n!\,(c)_n$, $g = \sum b_n x^n/n!\,(c)_n$.

2.4.1 Reproducing kernel

For the reproducing kernel, we introduce the function $\mathcal{I}_c$.

2.4.1.1 Definition. The function $\mathcal{I}_c$ is defined by the series

$$\mathcal{I}_c(x) = \sum_{n=0}^{\infty} \frac{x^n}{n!\,(c)_n}$$

This is a modified Bessel function. From Watson[47], p. 365, we have the definition of the Bessel function I_ν:

$$I_\nu(x) = \sum_{n=0}^{\infty} \frac{(x/2)^{\nu+2n}}{n!\,\Gamma(\nu+n+1)}$$

We note

2.4.1.2 Proposition. *Some properties of $\mathcal{I}_c$:*

1. *As a hypergeometric function*

$$\mathcal{I}_c(x) = {}_0F_1\left(\begin{array}{c} - \\ c \end{array} \middle|\, x\right)$$

2. *$\mathcal{I}_c$ can be expressed in terms of the Bessel function I_ν:*

$$\mathcal{I}_c(x) = \Gamma(\nu+1)\,x^{-\nu/2}I_\nu(2\sqrt{x}), \qquad x > 0$$

   where $\nu = c - 1$.

3. *$\mathcal{I}_c$ is an entire function satisfying the inequality*

$$\mathcal{I}_c(r) < e^{r/c}, \qquad r > 0$$

Proof: These follow readily from the definitions. For 3., note the inequality $c^n \le (c)_n$ for $n \ge 0$. ∎

Now we have

2.4.1.3 Proposition. *The reproducing kernel for Φ_c is*

$$\mathcal{I}_c(x\bar{y})$$

Proof: This follows from Prop. 2.4.1 and the defining equation, eq. (2.2.2.2).

∎

2.4.2 Operator calculus

As for the HW-type spaces, we can construct a calculus for the operator Δ. From Ch. 1, eq. (3.1.2), we have, for $2c$ a positive integer, the dimension of Euclidean space, an interesting interpretation of the sl(2) calculus in terms of 'one-half the Laplacian acting on functions of half the squared radius'. We have, with the notation Δ_β for the lowering operator:

2.4.2.1 Proposition. *For general β, with $\psi_n = x^n$, the lowering operator is realized as a second order differential operator:*

$$\Delta_\beta = \beta x \frac{d^2}{dx^2} + c \frac{d}{dx}$$

Proof: This follows from the definition of the action of L, as given in eq. (2.4.1), modified accordingly as in Ch. 1, Prop. 3.5.3. ∎

We consider here the standard sl(2) algebra with $\beta = 1$. Thus, the operator Δ is realized as the second order operator:

$$\Delta = x \frac{d^2}{dx^2} + c \frac{d}{dx}$$

Using the reproducing kernel as in §2.3, we have the following structure.

1. For a probability measure $p(dx)$, define the integral transform

$$\phi(x) = \int_{-\infty}^{\infty} \mathcal{I}_c(xy) \, p(dy)$$

   Prop. 2.4.1.2 suggests that this be called the *Hankel-Mellin* transform of the measure p (see §5.4 of Chapter 4 for more details).

2. The proof of Prop. 2.3.2.1 goes through since only the reproducing kernel property is used. Thus:

2.4.2.2 Proposition. *If the Hankel-Mellin transform $\phi(x)$ of the probability measure $p(dx)$ exists for all $x \in \mathbf{R}$, then $\phi \in \Phi_c$ if and only if $\int \phi(x) p(dx)$ is finite.*

3. We can interpret the Fock space inner product $\langle f, g \rangle$ in terms of Δ:

$$\langle f, g \rangle = f(\Delta) g(0) = \bar{g}(\Delta) f(0)$$

   Cf. Cor. 2.3.2.5.

4. The operator calculus is to be implemented via the Hankel-Mellin transform. First, we have

$$\langle f \rangle = \phi(\Delta) f(0)$$

with $\phi(x) = \int \mathcal{I}_c(xy) \, p(dy)$. However, the formulation of $\phi(\Delta) \, f(x)$ requires more considerations. The essential feature is to be found in the discussion above concerning $e^{yD} \, f(x)$. We wish to formulate the relation

$$\phi(\Delta) \, f(x) = \int_{-\infty}^{\infty} \mathcal{I}_c(y\Delta) f(x) \, p(dy)$$

For the HW case, we adopted the Def. 2.3.2.6. Recalling point 3. of the discussion concerning e^{yD} we note the following approach:

$$\phi(\Delta) \, f(a) = \langle \phi(\Delta) f(x), \mathcal{I}_c(ax) \rangle$$
$$= \int_{-\infty}^{\infty} \langle \mathcal{I}_c(y\Delta) f(x), \mathcal{I}_c(ax) \rangle \, p(dy)$$
$$= \int_{-\infty}^{\infty} \langle f(x), \mathcal{I}_c(yx) \mathcal{I}_c(ax) \rangle \, p(dy)$$

using the fact that here x and Δ are adjoints. The essential feature is

$$\mathcal{I}_c(y\Delta) \, f(a) = \langle f(x), \mathcal{I}_c(yx) \mathcal{I}_c(ax) \rangle$$

In the HW case, the group property of the exponential function comes in. Here we would employ a suitable *product formula* for Bessel functions (see §4.3). An alternative formulation is to find an integral representation for $\mathcal{I}_c(y\Delta) \, f(x)$ analogous to $e^{yD} \, f(x) = f(x + y)$, which we can write in this context in the form

$$e^{yD} \, f(x) = \int_{-\infty}^{\infty} f(x + s) \, \delta_y(ds)$$

This is the approach we take below.

Remark. The interplay of properties of the reproducing kernel Hilbert space and the group structure are features making this subject so interesting. There are many further aspects. See De Branges[12], De Branges-Rovnyak[13], and Vilenkin[46] for some topics related to these areas.

The *Gegenbauer polynomials* $C_m^\nu(\alpha)$, are defined via the generating function

$$(1 - 2\alpha v + v^2)^{-\nu} = \sum_{m=0}^{\infty} v^m C_m^\nu(\alpha)$$

We have explicitly:

2.4.2.3 Lemma. *The Gegenbauer polynomials have the form:*

$$C_m^\nu(\alpha) = \sum_s \frac{(-1)^{m+s}(\nu)_s}{(2s - m)! \, (m - s)!} \, (2\alpha)^{2s-m}$$

where the sum is over s such that $m \geq s \geq m/2$.

Proof: Expand via the binomial theorem:

$$\sum_{s=0}^{\infty} \frac{(\nu)_s}{s!} v^s (2\alpha - v)^s = \sum_{s=0}^{\infty} \frac{(\nu)_s}{s!} \binom{s}{\lambda} (2\alpha)^{\lambda} v^{2s-\lambda} (-1)^{s-\lambda}$$

using the summation convention. Setting $\lambda = 2s - m$ and rearranging gives the stated form. ∎

Next we have the matrix elements of the action of $\mathcal{I}_c(y\Delta)$ on the powers x^n.

2.4.2.4 Lemma. *The operator* $\mathcal{I}_c(y\Delta)$ *acts on* x^n *giving:*

$$\mathcal{I}_c(y\Delta)\, x^n = \sum_{k=0}^{n} \binom{n}{k} \frac{(c)_n}{(c)_{n-k}\,(c)_k}\, x^{n-k} y^k$$

Proof: This follows directly from Def. 2.4.1.1. Cf., matrix elements given in Ch. 1, Prop. 3.3.2.1. ∎

Remark. In the remainder of this section we require $c > \frac{1}{2}$.

Now we have the main lemma, expressing the matrix elements as an integral.

2.4.2.5 Lemma. *With B denoting the beta function,*

$$\binom{n}{k} \frac{(c)_n}{(c)_{n-k}\,(c)_k} = \frac{1}{2B(c - \frac{1}{2}, \frac{1}{2})} \int_0^{2\pi} C_{2k}^{-n}(\cos\phi)\,(\sin^2\phi)^{c-1}\, d\phi$$

Proof: By Lemma 2.4.2.3,

$$\int_0^{2\pi} C_{2k}^{-n}(\cos\phi)\,(\sin^2\phi)^{c-1}\, d\phi$$

$$= \sum_s \frac{(-1)^s(-n)_s}{(2s-2k)!\,(2k-s)!}\, 4^{s-k} \int_0^{2\pi} (\cos^2\phi)^{s-k}(\sin^2\phi)^{c-1}\, d\phi$$

By Prop. 3.3 of the Introduction, we have, noting that the factors of 4 cancel:

$$2\pi \sum_s \frac{(-1)^s(-n)_s}{(2s-2k)!\,(2k-s)!} \frac{(1/2)_{c-1}}{\Gamma(s-k+c)} \frac{(2s-2k)!}{(s-k)!}$$

$$= \left(\frac{1}{2}\right)_{c-1} 2\pi \sum_m \frac{(-1)^{m+k}(-n)_{m+k}}{(k-m)!\,\Gamma(m+c)\,m!}$$

where, since $s - k$ is an integer, we have used Intro., eq. (3.1), and made the substitution $s = k + m$. This last sum may be written

$$\frac{(n-k)!}{\Gamma(c+n-k)} \binom{n}{k} \sum_m \binom{k}{m} \binom{n-k+c-1}{n-k-m}$$

The summation is a Chu-Vandermonde sum, Ch.2, Prop. 4.1.1, equal to $\binom{n+c-1}{n-k}$. Taking out the normalization factor, $2\pi(1/2)_{c-1}/\Gamma(c) = 2B(c - 1/2, 1/2)$, and combining the remaining factors yields the result. ∎

Now we can write the action of $\mathcal{I}_c(y\Delta)$ analogously to that of the translation by y produced by the operator $\exp(yD)$.

2.4.2.6 Theorem. *The action of $\mathcal{I}_c(y\Delta)$ as an integral operator:*

$$\mathcal{I}_c(y\Delta)\,f(x) = \frac{1}{2B(c - \frac{1}{2}, \frac{1}{2})} \int_0^{2\pi} f(x + y - 2\sqrt{xy}\cos\phi)\,(\sin^2\phi)^{c-1}\,d\phi$$

Proof: For $f(x) = x^n$, with $t = \sqrt{x/y}$,

$$(x + y - 2\sqrt{xy}\cos\phi)^n = y^n(1 - 2t\cos\phi + t^2)^n$$

$$= y^n \sum_m t^m C_m^{-n}(\cos\phi)$$

$$= \sum_m x^{m/2}y^{n-m/2}C_m^{-n}(\cos\phi)$$

From Lemma 2.4.2.3, we see that if m is odd, only odd powers of $\cos\phi$ arise, yielding zero upon integration. Hence we may replace m by $2k$ in the sum and the result follows from Lemmas 2.4.2.4 and 2.4.2.5. ∎

Remark. From the proof of this Theorem, we see that the integral is an even function of $\sqrt{xy}$ and that the interpretation of $(\sqrt{xy})^{2k}$ is as $(xy)^k$, keeping signs of x, y in case they take negative values. For general $f \in C(\mathbf{R})$ we can restrict to $x, y > 0$.

We have now the following theorem/definition, cf. Def. 2.3.2.6. (Recall Prop. 2.4.2.2 and surrounding discussion.)

2.4.2.7 Theorem. *Given a probability measure $p(dx)$, let $\phi(x)$ denote its Hankel-Mellin transform. Then the operator $\phi(\Delta)$ is given by*

$$\phi(\Delta)\,f(x) = \frac{1}{2B(c - \frac{1}{2}, \frac{1}{2})} \int_{-\infty}^{\infty}\int_0^{2\pi} f(x + y - 2\sqrt{xy}\cos\vartheta)\,(\sin^2\vartheta)^{c-1}\,d\vartheta\,p(dy)$$

III. Tensor products

The tensor product space corresponds to a Fock space of two independent variables x_1, x_2, say, corresponding to two mutually commuting algebras, $\mathcal{A}_1$, $\mathcal{A}_2$. For definiteness, let us consider the sl(2) case. We have two algebras, with corresponding generators $\{L_1, R_1, \rho_1\}$ and $\{L_2, R_2, \rho_2\}$. We take as basis

$$\phi_{jk} = R_1^j R_2^k \Omega_{12}$$

where the vacuum Ω_{12} is thought of as the product of the two vacuums for the individual spaces. That is,

$$L_1\Omega_{12} = L_2\Omega_{12} = 0, \qquad \rho_1\Omega_{12} = c_1\Omega_{12}, \quad \rho_2\Omega_{12} = c_2\Omega_{12}$$

The combined algebra has generators

$$L = L_1 + L_2, \qquad R = R_1 + R_2, \qquad \rho = \rho_1 + \rho_2$$

which again satisfy the commutation relations for the sl(2) algebra. We want a vacuum Ω_n, constructed as a linear combination of the ϕ_{jk}, that satisfies $L\Omega_n = 0$ and $\rho\Omega_n = (c_1 + c_2 + 2n)\Omega_n$. I.e., Ω_n is homogeneous of degree n as a function of R_1, R_2. Then one has $\psi_{mn} = R^m\Omega_n$ as a basis for a representation of the sl(2) algebra corresponding to the operators $\{ L, R, \rho \}$. The *Clebsch-Gordan coefficients* are the matrix elements of the change of basis from the corresponding normalized states $\tilde{\phi}_{jk}$ to the orthonormal states $\tilde{\psi}_{mn}$, the normalized ψ_{mn}. We will see that these matrix elements yield a particular class of orthogonal polynomials associated to each algebra.

First we examine the HW case to see clearly what is going on.

3.1 HW AND KRAWTCHOUK POLYNOMIALS

We denote the operators R and L as usual by x and d. For the HW tensor product, we have as basis the monomials

$$\phi_{jk} = x_1^j x_2^k$$

Taking, e.g., $d_1 = a\,\partial/\partial x_1$ and $d_2 = b\,\partial/\partial x_2$, gives $[d_1, x_1] = a$, $[d_2, x_2] = b$, with $[d_1, d_2] = [x_1, x_2] = 0$. Now $d = d_1 + d_2$ and $x = x_1 + x_2$ act on the tensor product space. We have $[d, x] = a + b$. The first step is to find a vacuum state Ω_n that satisfies:

1. It is homogeneous of degree n in (x_1, x_2)
2. It is annihilated by d: $d\Omega = 0$

Then we have a HW representation space with basis $\psi_{mn} = x^m\Omega_n$. One readily sees

3.1.1 Proposition.   *For any $n \geq 0$, a vacuum state is given by*

$$\Omega_n = (bx_1 - ax_2)^n$$

Thus, for given n, a basis for a HW representation is given by

$$\psi_{mn} = (x_1 + x_2)^m (bx_1 - ax_2)^n$$

Furthermore, as m, n vary over the nonnegative integers, we have a basis for polynomials in the variables (x_1, x_2).

Now introduce the Fock space inner product. The two-variable space is the tensor product of the one-variable spaces and the norm is the product of the one-variable norms. I.e., recalling Prop. 2.3.1,

$$\|\phi_{jk}\|^2 = \|x_1^j x_2^k\|^2 = a^j j!\, b^k k! \tag{3.1.1}$$

Next we need the norms of the basis ψ_{mn}.

3.1.2 Proposition.   *The squared norms for the ψ_{mn} are*

$$\|\psi_{mn}\|^2 = n!\, m!\, (ab)^n (a+b)^{n+m}$$

Proof: For fixed m, ψ_{mn}, is a basis in the HW Fock space with raising operator x with $[d, x] = a + b$. So

$$\|\psi_{mn}\|^2 = \langle x^m \Omega_n, x^m \Omega_n \rangle = (a+b)^m m! \, \|\Omega_n\|^2$$

cf., Prop. 2.3.1. We compute the squared norm of Ω_n in the two-variable space:

$$\|\Omega_n\|^2 = \| \sum_j \binom{n}{j} (-1)^j b^{n-j} a^j x_1^{n-j} x_2^j \|^2$$

$$= \sum_j \binom{n}{j}^2 b^{2(n-j)} a^{2j} \| x_1^{n-j} x_2^j \|^2$$

$$= n!\, (ab)^n (a+b)^n$$

using the orthogonality and eq. (3.1.1).   ∎

3.1.3 Definition.   Define the coefficients $\langle\langle \cdot \,|\, \cdot \rangle\rangle$ for the change-of-basis by the relation:

$$\psi_{mn} = \sum_{j,k} \langle\langle mn | jk \rangle\rangle \, \phi_{jk}$$

Then we have

3.1.4 Proposition.   *For $N = n + m$, the $\langle\langle \cdot \,|\, \cdot \rangle\rangle$ coefficients are given by*

$$\langle\langle mn | N - k, k \rangle\rangle = b^n \binom{N}{k} {}_2F_1 \left(\begin{array}{c} -n, -k \\ -N \end{array} \,\Big|\, 1 + \frac{a}{b} \right)$$

where $\langle\langle mn | jk \rangle\rangle = 0$, if $j + k \neq N$.

Proof: Using Chapter 2, Prop. 2.1, set $N = n + m$, $t = x_2 / x_1$:

$$\psi_{mn} = (x_1 + x_2)^m (bx_1 - ax_2)^n = b^n x_1^N (1 + t)^m (1 - (a/b)t)^n$$

$$= b^n x_1^N \sum \binom{N}{k} t^k \, {}_2F_1 \left(\begin{array}{c} -n, -k \\ -N \end{array} \,\Big|\, 1 + \frac{a}{b} \right)$$

and the result follows after multiplying x_1^N back in.   ∎

Now we denote by $\tilde{\psi}_{mn}$ and $\tilde{\phi}_{jk}$ the *normalized states*:

$$\tilde{\phi}_{jk} = \phi_{jk} / \|\phi_{jk}\|, \qquad \tilde{\psi}_{mn} = \psi_{mn} / \|\psi_{mn}\|$$

3.1.5 Definition.   The *CG (Clebsch-Gordan) coefficients* $< \cdot \,|\, \cdot >$ are defined by the relations

$$\tilde{\psi}_{N-n,n} = \sum_k \langle Nn | k \rangle \, \tilde{\phi}_{N-k,k}$$

Combining the above results, we have

3.1.6 Theorem.   *The CG coefficients for the HW algebra are, for $0 \leq k, n \leq N$:*

$$\langle Nn|k \rangle = \left[\binom{N}{n} \binom{N}{k} \frac{a^{N-n-k} b^{n+k}}{(a+b)^N} \right]^{1/2} {}_2F_1 \left(\begin{array}{c} -k, -n \\ -N \end{array} \middle| \; 1 + \frac{a}{b} \right)$$

Proof: It remains to put in the norm factors in Prop. 3.1.4, as it is clear from the definitions that

$$\langle Nn|k \rangle = \langle \langle N-n, n|N-k, k \rangle \rangle \times \|\phi_{N-k,k}\| / \|\psi_{N-n,n}\|$$

Check that the coefficient of the ${}_2F_1$

$$\left[\binom{N}{k}^2 b^{2n} \frac{a^{N-k} b^k (N-k)!\, k!}{n!\,(N-n)!\,(ab)^n (a+b)^N} \right]^{1/2}$$

reduces as indicated.   ∎

3.1.1 Krawtchouk polynomials

The fact that we are transforming from one orthonormal set to another means that the CG coefficients form an *orthogonal matrix*.

3.1.1.1 Theorem.   *For fixed N:*

$$\sum_k \langle Nn|k \rangle \langle Nn'|k \rangle = \delta_{nn'}$$

$$\sum_n \langle Nn|k \rangle \langle Nn|k' \rangle = \delta_{kk'}$$

Proof: The statement is saying that the matrix $P_{nk} = \langle Nn|k \rangle$, is orthogonal. I.e., with $P^\dagger$ denoting the transpose of P, $PP^\dagger = P^\dagger P = I$. This follows from orthonormality of the states. E.g.,

$$\delta_{nn'} = \langle \tilde{\psi}_{N-n,n}, \tilde{\psi}_{N-n',n'} \rangle = \sum_k \sum_{k'} P_{nk} P_{n'k'} \langle \tilde{\phi}_{N-k,k}, \tilde{\phi}_{N-k',k'} \rangle$$

$$= \sum_k P_{nk} P_{n'k}$$

■

In the present context,

3.1.1.2 Theorem. *The orthogonality relations have the form:*

$$\delta_{kk'} = \left[\binom{N}{k} \binom{N}{k'} \right]^{1/2} \left(\frac{b}{a} \right)^{(k+k')/2} \times$$

$$\sum_n \binom{N}{n} \frac{a^{N-n} b^n}{(a+b)^N} {}_2F_1 \left(\begin{array}{c} -n, -k \\ -N \end{array} \middle| 1 + \frac{a}{b} \right) {}_2F_1 \left(\begin{array}{c} -n, -k' \\ -N \end{array} \middle| 1 + \frac{a}{b} \right)$$

and similarly for the sum over k for given N, n, n'.

We remark some features of interest. First note the symmetry between the variables n and k. Thus we can focus on one of these variables, while corresponding results will hold for the other. Observe that for N and $k \leq N$ fixed, $\langle Nn|k \rangle$ gives a polynomial of degree k in the variable n.

Now define parameters

$$p = \frac{a}{a+b}, \qquad q = \frac{b}{a+b}$$

which satisfy $0 < p, q < 1$ and $p + q = 1$, since $a, b > 0$. We have the *binomial probability distribution* given by

$$\binom{N}{n} p^n q^{N-n}, \qquad 0 \leq n \leq N$$

3.1.1.3 Definition. Define the *Krawtchouk polynomials* $K_k(x, N)$, by

$$K_k(x, N) = N^{(k)} q^k {}_2F_1 \left(\begin{array}{c} -k, -x \\ -N \end{array} \middle| \frac{1}{q} \right)$$

Then we have

3.1.1.4 Theorem. *The Krawtchouk polynomials are orthogonal with respect to the binomial distribution:*

$$\sum_n \binom{N}{n} p^{N-n} q^n K_k(n, N) K_{k'}(n, N) = k! \, N^{(k)} (pq)^k \delta_{kk'}$$

Proof: This follows directly from Theorem 3.1.1.2 by substituting in the Krawtchouk polynomials. Observe that the factor a/b is the same as p/q. ∎

Remark. The relation $\tilde{\psi}_{N-n,n} = \sum \langle Nn|k \rangle \, \tilde{\phi}_{N-k,k}$ is thus interpreted as a generating function for the Krawtchouk polynomials.

3.2 sl(2) AND HAHN POLYNOMIALS

We can make a similar study for the sl(2) algebra. Take two sl(2)-type Fock spaces with orthogonal bases $x_1^j = R_1^j \Omega_1$ and $x_2^k = R_2^k \Omega_2$ and form the tensor product space with basis $\phi_{jk} = x_1^j x_2^k = R_1^j R_2^k \Omega_{12}$. The squared norms are given by, Prop. 2.4.1:

$$\gamma_j(1) = \|x_1^j\|^2 = j!(c_1)_j, \qquad \gamma_k(2) = \|x_2^k\|^2 = k!(c_2)_k$$

with

$$\|\phi_{jk}\|^2 = \gamma_j(1)\gamma_k(2) \qquad (3.2.1)$$

The algebra acting on the product space is generated by $L = L_1 + L_2$, $R = R_1 + R_2$, and $\rho = \rho_1 + \rho_2$, a basis for a standard sl(2) representation. On the vacuum states, setting $c = c_1 + c_2$:

$$\rho_1 \Omega_{12} = c_1 \Omega_{12}, \qquad \rho_2 \Omega_{12} = c_2 \Omega_{12}, \qquad \rho \Omega_n = (c + 2n)\Omega_n$$

that is, Ω_n is homogeneous of degree n in (x_1, x_2). Explicitly,

3.2.1 Proposition. *The function*

$$\Omega_n = \sum_k \frac{(-1)^k x_1^{n-k} x_2^k}{(n-k)!\,k!\,(c_1)_{n-k}(c_2)_k}$$

is a vacuum state for the tensor product representation.

Proof: Write this in the form

$$\Omega_n = \sum_k \frac{(-1)^k x_1^{n-k} x_2^k}{\gamma_{n-k}(1)\,\gamma_k(2)}$$

The action of L_1, e.g., is

$$L_1 x_1^j = j(c_1 + j - 1)\,x_1^{j-1}$$

and similarly for L_2 on x_2^k, so that on applying L, each index is shifted by one:

$$L\Omega_n = \sum_k (-1)^k \left[\frac{x_1^{n-k-1} x_2^k}{\gamma_{n-k-1}(1)\,\gamma_k(2)} + \frac{x_1^{n-k} x_2^{k-1}}{\gamma_{n-k}(1)\,\gamma_{k-1}(2)} \right]$$

$$= \sum_k \left[(-1)^k \frac{x_1^{n-k-1} x_2^k}{\gamma_{n-k-1}(1)\,\gamma_k(2)} + (-1)^{k+1} \frac{x_1^{n-k-1} x_2^k}{\gamma_{n-k-1}(1)\,\gamma_k(2)} \right]$$

$$= 0$$

Remark. We can express this in terms of a hypergeometric function:

$$\Omega_n = \frac{x_1^n}{n!(c_1)_n} \, {}_2F_1 \left(\begin{matrix} -n, 1 - c_1 - n \\ c_2 \end{matrix} \, \middle| \, -\frac{x_2}{x_1} \right)$$

using the relation, cf. Introduction, eq. (3.2):

$$(A)_{B-k} = \frac{(-1)^k (A)_B}{(1 - A - B)_k}$$

with $A = c_1$, $B = n$.

The vacuum state can be written via *Jacobi polynomials* (see reference list in Intro.)

$$\Omega_n = \frac{(-1)^n (x_1 + x_2)^n}{(c_1)_n (c_2)_n} \, P_n^{(c_1 - 1, c_2 - 1)}((x_2 - x_1)/(x_2 + x_1))$$

We note as well the relation which will prove useful in the following, cf. Introduction, eqs. (3.2),(3.3):

$$\frac{(A)_n}{(A)_{n-k}} = \binom{A + n - 1}{k} k! \tag{3.2.2}$$

which follows by converting from Γ-functions to factorials.

Next we need the norm of Ω_n:

3.2.2 Proposition.   *The squared norm of the vacuum state is*

$$\|\Omega_n\|^2 = \frac{1}{(c_1)_n (c_2)_n} \binom{c + 2n - 2}{n}$$

Proof: Write

$$\Omega_n = \sum_k \frac{(-1)^k \phi_{n-k,k}}{\gamma_{n-k}(1) \gamma_k(2)}$$

with the ϕ's orthogonal, so that

$$\|\Omega_n\|^2 = \sum_k \frac{\gamma_{n-k}(1) \gamma_k(2)}{(\gamma_{n-k}(1) \gamma_k(2))^2}$$

$$= \sum_k \frac{1}{(n - k)! \, k! \, (c_1)_{n-k} (c_2)_k}$$

$$= \frac{1}{(c_1)_n (c_2)_n} \sum_k \binom{c_1 + n - 1}{k} \binom{c_2 + n - 1}{n - k}$$

using eq. (3.2.2). Now the result follows via the Chu-Vandermonde sum, Ch. 2, Prop. 4.1.1.   ∎

We construct the basis

$$\psi_{mn} = R^m \Omega_n = (x_1 + x_2)^m \Omega_n$$

Acting on this basis, we have a representation of the standard sl(2) algebra, generated by $\{L, R, \rho\}$, as noted above.

3.2.3 Proposition.   *The squared norms are given by:*

$$\|\psi_{mn}\|^2 = \frac{m!\,(c+2n)_m}{(c_1)_n(c_2)_n}\binom{c+2n-2}{n}$$

Proof: Since $\rho\Omega_n = (c+2n)\Omega_n$, the squared norm is

$$\|\psi_{mn}\|^2 = \|R^m\Omega_n\|^2 = m!(c+2n)_m\,\|\Omega_n\|^2$$

and the result follows from Prop. 3.2.2.   ■

Denoting the coefficients for the unnormalized bases by double brackets, Def. 3.1.3:

$$\psi_{mn} = \sum_{j,k}\langle\!\langle mn|jk\rangle\!\rangle\phi_{jk}$$

we have

3.2.4 Proposition.   *For sl(2), the coefficients $\langle\!\langle\,\cdot\,|\,\cdot\,\rangle\!\rangle$ are given by:*

$$\langle\!\langle mn|jk\rangle\!\rangle = \sum_s \binom{m}{k-s}\frac{(-1)^s}{(c_1)_{n-s}(c_2)_s\,(n-s)!\,s!}$$

or, equivalently,

$$\langle\!\langle mn|jk\rangle\!\rangle = \frac{1}{(c_1)_n(c_2)_n}\sum_s \binom{m}{k-s}\binom{c_1+n-1}{s}\binom{c_2+n-1}{n-s}(-1)^s$$

where the coefficient vanishes unless $j = m+n-k$.

Proof: Via Prop. 3.2.1, we have, using our summation convention:

$$\psi_{mn} = \sum_s \binom{m}{\lambda}x_1^{m-\lambda}x_2^{\lambda}\frac{(-1)^s x_1^{n-s}x_2^s}{(c_1)_{n-s}(c_2)_s\,(n-s)!\,s!}$$

$$= \sum_s \phi_{m+n-\lambda-s,\lambda+s}\binom{m}{\lambda}\frac{(-1)^s}{(c_1)_{n-s}(c_2)_s\,(n-s)!\,s!}$$

Setting $k = \lambda+s$, $j = m+n-\lambda-s$, and the first part of the result follows. For the second form, use eq. (3.2.2) to convert to binomial coefficients.   ■

We have now

3.2.5 Theorem.   *For the sl(2) algebra, the CG coefficients are given by*

$$\langle Nn|k\rangle = \left[\binom{c+2n-2}{n}\binom{c+n+N-1}{N-n}(c_1)_n(c_2)_n\right]^{-1/2}\left[\frac{(c_1)_{N-k}(c_2)_k}{(N-k)!\,k!}\right]^{1/2}$$

$$\times \sum_s \binom{N-k}{n-s}\binom{k}{s}\binom{c_1+n-1}{s}\binom{c_2+n-1}{n-s}s!(n-s)!(-1)^s$$

Proof: We have $\langle Nn|k\rangle = \langle\langle mn|jk\rangle\rangle \times \|\phi_{mn}\|/\|\psi_{mn}\|$, with $m+n = j+k = N$. Thus, putting in appropriate norm factors, eq. (3.2.1)and Prop. 3.2.3, and using additional factors $(c_1)_n(c_2)_n$, gives

$$\left[\frac{(N-k)!\,k!}{m!\,(c+2n)_m}\binom{c+2n-2}{n}^{-1}\frac{(c_1)_{N-k}(c_2)_k}{(c_1)_n(c_2)_n}\right]^{1/2}\times$$

$$\sum_s \binom{m}{k-s}\frac{(-1)^s(c_1)_n(c_2)_n}{(c_1)_{n-s}(c_2)_s(n-s)!\,s!}$$

With $m = N - n$, observe that

$$\binom{m}{k-s}\frac{(N-k)!\,k!}{(n-s)!\,s!} = m!\binom{N-k}{n-s}\binom{k}{s}$$

Multiplying and dividing by $(N-k)!k!$ and pulling out $m!$ yields:

$$\left[\binom{c+2n-2}{n}^{-1}\frac{m!}{(c+2n)_m}\frac{1}{(c_1)_n(c_2)_n}\frac{(c_1)_{N-k}(c_2)_k}{(N-k)!\,k!}\right]^{1/2}$$

$$\times \sum_s \binom{N-k}{n-s}\binom{k}{s}\frac{(c_1)_n(c_2)_n}{(c_1)_{n-s}(c_2)_s}(-1)^s$$

And rewriting

$$\frac{m!}{(c+2n)_m} = \binom{c+2n+m-1}{m}^{-1} = \binom{c+n+N-1}{N-n}^{-1}$$

gives the desired result.   ■

Remark. The summation term may be expressed in the form

$$\sum_s \binom{N-k}{n-s}\binom{k}{s}\frac{(c_1)_n(c_2)_n}{(c_1)_{n-s}(c_2)_s}(-1)^s \qquad (3.2.3)$$

3.2.1 Hahn polynomials

We see that the summation, the expression in eq. (3.2.3), is a polynomial in k of degree n. Thus

3.2.1.1 Definition.   For given N, c_1, c_2, the *Hahn polynomials* are given by

$$Ha_n(x) = \sum_s \binom{N-x}{n-s}\binom{x}{s}\binom{c_1+n-1}{s}\binom{c_2+n-1}{n-s} s!(n-s)!(-1)^s$$

Next, we bring in the *hypergeometric distribution* given by

$$p(N,a,b;k) = \binom{a}{N-k}\binom{b}{k} \Big/ \binom{a+b}{N} \qquad (3.2.1.1)$$

for positive integers a, b, where $0 \le k \le N$. Note that the orthogonality relation, Theorem 3.1.1.1, $\sum \langle Nn|k\rangle\langle Nn'|k\rangle = \delta_{nn'}$ may be written as a polynomial identity in the variables c_1, c_2. Thus we may extend it to arbitrary values of c_1, c_2, in particular for negative integers (as long as we avoid vanishing denominators in the resulting formulas). Recalling Theorem 3.1.1.4, we have similarly,

3.2.1.2 Theorem.   *The Hahn polynomials are orthogonal with respect to the hypergeometric distribution:*

$$\sum_k p(N,-c_1,-c_2;k)\,Ha_n(k)Ha_{n'}(k) = \frac{(c_1)_n\,(c_2)_n}{c+2n-1}\binom{N}{n}\frac{(c+n-1)_{N+1}}{(c)_N}$$

where c_1 and c_2 are negative integers, and the probabilities $p(N,-c_1,-c_2;k)$ are as in eq. (3.2.1.1).

Proof: Comparing with Theorem 3.2.5, we rewrite

$$\frac{(c_1)_{N-k}}{(N-k)!}\frac{(c_2)_k}{k!} = \frac{(c)_N}{N!}\binom{-c_1}{N-k}\binom{-c_2}{k}\Big/\binom{-c}{N}$$

$$= \binom{c+N-1}{N}\times p(N,-c_1,-c_2;k)$$

and after moving the normalizing factors over, the right hand side becomes

$$\delta_{nn'}\,(c_1)_n(c_2)_n\binom{c+2n-2}{n}\binom{c+n+N-1}{N-n}\Big/\binom{c+N-1}{N}$$

Rewriting in terms of $\Gamma-$functions yields the result stated.   ∎

Note that we can also write the squared norms in the form:

$$\frac{(c_1)_n\,(c_2)_n}{c+2n-1}\binom{N}{n}\frac{(c+N)_n}{(c)_{n-1}}$$

Remark. For background on quantum theory, see Bohm[7], Landau&Lifshitz[31], Biedenharn&Louck[6]. There is much in these books related to the material in this chapter. For material on integral transforms and connections with harmonic analysis, see Lebedev[32] and Helgason[25]. For background on probability theory, see Feller[20], Breiman[10].

IV. Exercises and examples

4.1 EXERCISES

1. a. Show that the Pauli matrices, §1.1.1, are a basis for Hermitian 2×2 matrices.

   b. Show by direct calculation that they are orthogonal with respect to the inner product $\langle X, Y \rangle = \operatorname{tr}(XY^*)$ as claimed in Proposition 1.1.1.1.

   c. Verify the commutation and anticommutation relations stated in Proposition 1.1.1.1.

2. a. Show that if A is a Hermitian matrix, then e^{iA} is unitary.

   b. In particular, calculate $\exp(i\sigma_k)$ for the Pauli matrices σ_k, $k = 1, 2, 3$, and check unitarity explicitly.

3. Referring to §2.1:

   a. Verify that $[V, R] = I$ acting on the basis ψ_n.

   b. Find $[L, R]$ and $[L, [L, R]]$.

   c. Give a formula for $LR^n\Omega$.

4. Referring to Proposition 2.2.2.1, derive the bound

$$|K(x, y)| \le \sqrt{K(x, x)}\sqrt{K(y, y)}$$

5. In §2.3.2, show that for $x \in \mathbf{R}$, the transform $\phi(x) = \int e^{xy} p(dy)$ is a positive convex function.

6. Prove the claim that the Fock spaces Φ_h are nested, i.e., $h > h'$ implies $\Phi_h \subset \Phi_{h'}$.

7. a. Show that the Fourier-Laplace transform of the Gaussian distribution with density

$$p(x) = e^{-x^2/2t} / \sqrt{2\pi t}, \qquad t > 0$$

   satisfies the condition of Proposition 2.3.2.1 only if $t < 1$.

   b. Discuss the gamma density $(t > 0)$

$$p(x) = x^{t-1}e^{-x} / \Gamma(t), \qquad x \ge 0$$

   in the context of Proposition 2.3.2.1.

8. In the discussion of point #1 following Definition 2.3.2.6, check the formula

$$e^{yD} f(x) = f(x + y) e^{yD}$$

using the HW commutation rules (Chapter 1).

9. a. For polynomials $f(x) = \sum a_n x^n$, $g(x) = \sum b_n x^n$, verify Corollary 2.3.2.5 by direct calculation.

b. Extending the Corollary to power series find $\langle \sinh x, x^n \rangle$ and $\langle \cosh x, x^n \rangle$. Check consistency with $\cosh x = (e^x + e^{-x})/2$ and similarly for $\sinh x$. Use $\exp(\pm D)x^n = (x \pm 1)^n$.

10. Check the details of Proposition 2.4.1.

11. Fill in the details of Proposition 2.4.1.2. Discuss Proposition 2.4.1.3 analogously to Propositions 2.3.3 (cf. Proposition 2.2.2.1) and 2.3.1.1 and their proofs.

12. Fill in the details of the proof of Lemma 2.4.2.3.

13. a. Verify the statement in §III that for the tensor product the operators $L = L_1 + L_2$, $R = R_1 + R_2$, $\rho = \rho_1 + \rho_2$ indeed satisfy the sl(2) commutation relations.

b. Formulate the tensor product of n copies of a Lie algebra, specifically for sl(2).

14. Check Theorem 3.1.1.4 as a consequence of Theorem 3.1.1.2. Verify explicitly for $N = 2, 3$.

15. Verify the remark following Proposition 3.2.1 expressing the vacuum Ω_n as a hypergeometric function.

4.2 AN INDEFINITE FORM ON MATRICES

Let $E_0 = I$, $E_1 = \begin{pmatrix} 0 & 1 \\ 1 & 0 \end{pmatrix}$, $E_2 = \begin{pmatrix} 0 & 1 \\ -1 & 0 \end{pmatrix}$ and $E_3 = \begin{pmatrix} 1 & 0 \\ 0 & -1 \end{pmatrix}$.

1. Show that they satisfy ($i, j = 1, 2, 3$)

a. $E_i E_j = -E_j E_i$, for $i \neq j$.

b. $E_1 E_2 = -E_3$, $E_1 E_3 = -E_2$, $E_2 E_3 = -E_1$

c. $E_1^2 = E_3^2 = I$, $E_2^2 = -I$.

2. They are an orthogonal basis for the symmetric bilinear form $\langle X, Y \rangle = \text{tr}(XY)$. Notice that this is not positive definite. From the E's, we have the signature $+-++$, the minus coming from $E_2^2 = -I$.

3. Let l_A and r_A denote the operators of left and right multiplication by A respectively. These operators are adjoint with respect to the form $\langle X, Y \rangle$:

$$\langle l_A X, Y \rangle = \langle X, r_A Y \rangle$$

Thus the operator $\alpha = l_A + r_A$ is symmetric with respect to this form.

4. Can we diagonalize the operator α? Let $A = \left(\begin{smallmatrix} a & b \\ c & d \end{smallmatrix} \right)$. Show that the expansion of A in terms of the E's is $A = \sum c_i E_i$ where

$$c_0 = \tfrac{1}{2}(a+d), \qquad c_1 = \tfrac{1}{2}(b+c), \qquad c_2 = \tfrac{1}{2}(b-c), \qquad c_3 = \tfrac{1}{2}(a-d)$$

5. We are looking for matrices X such that $\alpha X = AX + XA = \lambda X$, eigenvectors of the operator α with eigenvalue λ. From the expansion in E's find a two-dimensional subspace of solutions as follows. Consider the space of X's orthogonal to the span of $\{A, I\}$ with respect to the form $\langle \, , \, \rangle$.

a. Such an X has an expansion $X = uE_1 + vE_2 + wE_3$, say. Show that

$$\tfrac{1}{2}\langle X, A \rangle = uc_1 - vc_2 + wc_3$$

b. Show that taking $u = 0$, $v = c_3$, $w = c_2$ and $u = c_2$, $v = c_1$, $w = 0$, gives two solutions.

c. Now take $X = c_3 E_2 + c_2 E_3$. Compute

$$AX + XA = 2(c_0 X - c_2 c_3 I + c_3 c_2 I) = (2c_0)X$$

Thus X is a solution with eigenvalue $2c_0 = \operatorname{tr}(A)$.

d. Similarly for $X = c_2 E_1 + c_1 E_2$.

6. What about other possibilities? Recall that, according to the Cayley-Hamilton theorem, A satisfies a characteristic equation: $(A - \lambda)(A - \mu) = 0$, where λ, μ are the eigenvalues of A. Notice that since A is arbitrary, these are in general complex numbers. Rewrite the characteristic equation in the form

$$(A - \lambda)A = \mu(A - \lambda) = A(A - \lambda)$$

and set $X = A - \lambda$ to find

$$AX + XA = 2\mu X$$

and similarly for $A - \mu$. These thus give the sought solutions with eigenvalues 2λ, 2μ for the operator α.

In summary, the operator $\alpha(X) = AX + XA$ is diagonalizable. Its spectrum consists of 2 times the spectrum of A and $\mathrm{tr}\,(A)$ with multiplicity two.

What about higher dimensions? We just state some results about orthogonal bases.

7. The bilinear form $\langle X, Y \rangle = \mathrm{tr}\,(XY)$ has signature consisting of $\binom{n+1}{2}$ +'s and $\binom{n}{2}$ −'s.

8. You can find an orthogonal basis consisting of $n^2 - n$ matrices of trace zero, $\binom{n}{2}$ of them skew-symmetric and an equal number of them symmetric. With the identity, this gives $n^2 - n + 1$. The remaining $n - 1$ can be chosen to be diagonal with all but one of the non-zero entries equal to 1.

9. For example, $n = 2$ gives the Minkowski metric, $n = 3$ has signature 6 +'s, 3 −'s, $n = 4$ has 10 +'s, 6 −'s, etc.

10. We close this section with the simple observation that if A is symmetric and B is skew-symmetric, then they are orthogonal: $\langle A, B \rangle = 0$.

4.3 PROPERTIES OF $\mathcal{I}_c$

Here we will look at some properties of $\mathcal{I}_c$, in particular, we will derive the product formula $\mathcal{I}_c$ for integer values of c by induction. Here we have $\Delta = x\,(d/dx)^2 + c\,(d/dx)$ and Δ_a the corresponding operator on functions of the variable a.

1. Show that:

   a. $\Delta \mathcal{I}_c(xy) = y\,\mathcal{I}_c(xy)$.

   b. $c\mathcal{I}_c'(x) = \mathcal{I}_{c+1}(x)$.

   c. $-x\mathcal{I}_{c+1} = c(c - 1)[\mathcal{I}_c(x) - \mathcal{I}_{c-1}(x)]$.

   Now, for the product formula, let

   $$\mathcal{I}_n(ax^2)\mathcal{I}_n(ay^2) = \Phi_n(a) = \int_0^{2\pi} \mathcal{I}_n(aw^2)C_n(\phi)\,d\phi/2\pi$$

where $w^2 = x^2 + y^2 - 2xy \cos \phi$.

2. Show that $C_1(\phi) = 1$, i.e.,

   $$\mathcal{I}_1(ax^2)\mathcal{I}_1(ay^2) = \int_0^{2\pi} \mathcal{I}_1(aw^2)\,d\phi/2\pi$$

3. From $\Delta_a(fg) = g\,\Delta_a f + f\,\Delta_a g + 2af'g'$ derive

   $$\Delta_a \Phi_n(a) = (x^2 + y^2)\Phi_n(a) + (2ax^2y^2/n^2)\Phi_{n+1}(a)$$

Comparing with

$$\Delta_a \Phi_n(a) = \int_0^{2\pi} \mathcal{I}_n(aw^2) w^2 C_n(\phi)\, d\phi/2\pi$$

find the recurrence formula

$$C_{n+1}(\phi) = 2n \sin\phi \int_0^\phi C_n(\phi') \cos\phi'\, d\phi'$$

4. Deduce

$$C_n(\phi) = 4^{n-1}(\sin^{2n-2}\phi)/\binom{2n-2}{n-1} = \frac{\Gamma(n)}{(1/2)_{n-1}} \sin^{2n-2}\phi$$

5. Write out the resulting product formula for $\mathcal{I}_n$.

Recall #4 of the discussion in §2.4.2, and compare with Theorem 2.4.2.6.

4.4 FORMULAS FOR GEGENBAUER POLYNOMIALS

Here are some expressions for Gegenbauer polynomials that are of interest.

1. In terms of hypergeometric functions:

$$C_n^\nu(\cos\theta) = e^{in\theta} \frac{(2\nu)_n}{n!} \,_2F_1\left(\begin{array}{c} -n,\nu \\ 2\nu \end{array} \middle| 1 - e^{-2i\theta}\right)$$

and

$$C_n^\nu(\cos\theta) = \frac{(\nu)_n}{n!} (2\cos\theta)^n \,_2F_1\left(\begin{array}{c} -n/2, (1-n)/2 \\ 1-\nu-n \end{array} \middle| \sec^2\theta\right)$$

2. For even and odd n we have:

If $n = 2p$ is even:

$$C_{2p}^\nu(\cos\theta) = \binom{-\nu}{p} \,_2F_1\left(\begin{array}{c} -p, \nu+p \\ \frac{1}{2} \end{array} \middle| \cos^2\theta\right)$$

If $n = 2p + 1$ is odd:

$$C_{2p+1}^\nu(\cos\theta) = \frac{(-1)^p(\nu)_{p+1}}{p!} (2\cos\theta) \,_2F_1\left(\begin{array}{c} -p, \nu+p+1 \\ \frac{3}{2} \end{array} \middle| \cos^2\theta\right)$$

See Rainville[40], Szëgo[45] for further details.

Chapter 4   MOMENT SYSTEMS

In this chapter, the emphasis is on the *vector space* structure. Then, in Chapter 5, we combine the theory of moment systems with the Fock space constructions of Chapter 3.

Here we will study some general constructions of representations of the HW and sl(2) algebras. There are two main techniques. For both HW and sl(2), starting with a basic realization, we apply a one-parameter group of automorphisms to produce representations, corresponding to *evolutions*, dynamical systems. For the HW algebra, we have additionally the technique of change-of-variables, §IV.

The moment systems are algebraically the same as x^n. As in the construction of representations in Chapters 1 and 3, we will construct bases for the vector space of polynomials. A given basis $\{\, h_n(x)\,\}_{n \geq 0}$ has an associated raising operator R, so that $h_n(x) = R^n h_0$, with $\deg h_n(x) = n$, for $n \geq 0$. Of particular interest is the analytic realization of the operator R and associated structures — lowering operator, generating function, and time-evolution.

I.   Moment generating functions and convolution

We are interested in stochastic processes of the type 'sum of independent, identically distributed random variables'. Now we discuss the basic features of this point of view.

1.1   RANDOM VARIABLES

We consider random variables corresponding to measures on spaces $\mathbf{R}^N$. For the approach via general probability spaces, we refer to Feller[20] and Breiman[10]. Here we offer a presentation in line with the algebraic approach of Chapter 3.

1.1.1 Definition.   The algebra of *bounded continuous functions* on $\mathbf{R}^N$ is denoted $\mathcal{C}(\mathbf{R}^N)$.

Given a probability measure $p(dx)$ on $\mathbf{R}$, we call the associated variable, i.e., the identity function on $\mathbf{R}$, denoted by X, the *random variable* with distribution $p(dx)$. Thus, the expectation of $f \in \mathcal{C}(\mathbf{R})$ determined by $p(dx)$ is denoted by $\langle f(X)\rangle$:

$$\langle f(X)\rangle = \int_{-\infty}^{\infty} f(x)\,p(dx)$$

So, X effectively denotes the variable of integration corresponding to the measure $p(dx)$.

Similarly, a family of random variables $(X_1, \ldots, X_N)$ corresponds to a measure $p(dx_1, \ldots, dx_N)$ on $\mathbf{R}^N$, with expectations defined on $\mathcal{C}(\mathbf{R}^N)$ by:

$$\langle f(X_1, \ldots, X_N) \rangle = \int_{\mathbf{R}^N} f(x_1, \ldots, x_N) \, p(dx_1, \ldots, dx_N)$$

An infinite family of random variables $\{X_t\}_{t \geq 0}$ with integer or real t is a *stochastic process* . Typically some regularity conditions are imposed. (See references below, §1.4.1.) Such a process corresponds to a probability measure on an infinite-dimensional space, a space of functions. In this work, we will only deal explicitly with finite-dimensional distributions.

Although the measure $p(dx)$ is determined by the expectations of functions in $\mathcal{C}(\mathbf{R})$, we are interested in cases where we can define $\langle h(X) \rangle = \int h(x) p(dx)$ for all polynomials $h(x)$. It is sufficient that the Fourier-Laplace transform, Ch. 3, §2.3.2,

$$\phi(a) = \int_{-\infty}^{\infty} e^{ax} \, p(dx)$$

exist for all $a \in \mathbf{R}$ in some interval containing the origin. In fact, we can extend to complex values of a, so that the integral exists in a neighborhood of the origin in $\mathbf{C}$. In such a case, the expectation is defined on the algebra of polynomials as well as on $\mathcal{C}(\mathbf{R})$. We thus adopt the following notations:

1.1.2 Definition. The algebra of polynomials on $\mathbf{R}$ is denoted by $\mathcal{P}$.

1.1.3 Definition. The set of probability measures that have Fourier-Laplace transform analytic in a neighborhood of the origin in $\mathbf{C}$ is denoted by $\mathcal{P}^*$.

1.2 MOMENTS AND TRANSFORMS

Let $p \in \mathcal{P}^*$. Denoting the corresponding random variable by X, we have the integral

$$\phi(z) = \langle e^{zX} \rangle = \int_{-\infty}^{\infty} e^{zx} \, p(dx)$$

analytic in z. Then we have the *moments* defined by

$$\mu_n = \langle X^n \rangle = \int_{-\infty}^{\infty} x^n \, p(dx)$$

and the expansion

$$\phi(z) = \sum_{n=0}^{\infty} \frac{z^n}{n!} \mu_n$$

Thus $\phi(z)$ is called the *moment generating function* . Since $\phi(0) = 1$ is nonzero at the origin, it may be expressed, near the origin, in the form of an exponential function. So we present the canonical form to be used throughout:

$$e^{H(z)} = \int_{-\infty}^{\infty} e^{zx}\, p(dx)$$

Typically, $\mu_1 = \langle X \rangle$, the *mean* of X, is denoted by μ. The *variance* $\sigma^2(X) = \langle (X - \mu)^2 \rangle$. We employ the terminology:

1.2.1 Definition. The random variable X is *centered* if $\mu = 0$. It is *scaled* when $\sigma^2(X) = 1$.

The scaling refers to multiplying X by a scale factor. We could say as well, e.g., 'X is scaled to variance 2', to indicate that we are specifying $\sigma^2(X) = 2$. Note that for any X for which μ_2 exists, $(X - \mu)/\sigma$ is centered and scaled. To relate to the moment generating function:

1.2.2 Proposition. *For the moment generating function of the form* $\exp(H(z))$

$$H(0) = 0, \qquad H'(0) = \mu, \qquad H''(0) = \sigma^2$$

Proof: That $H(0) = 0$ is the fact that the integral of $p(dx)$ is normalized to 1. The rest follows directly from the definitions. Use

$$\langle e^{zX} \rangle = e^{H(z)}$$

and differentiate under the expectation, which is valid in particular under the assumption that the moment generating function exists. ∎

Thus, if X is centered and scaled, $H(0) = H'(0) = 0$, and $H''(0) = 1$.

Remark. The *Fourier transform* of the measure $p(dx)$:

$$\phi(it) = \int_{-\infty}^{\infty} e^{itx}\, p(dx)$$

is called, after P. Lévy, the *characteristic function* of the distribution $p(dx)$. Note that knowledge of the characteristic function is the same as knowing expectations of the form $\langle \sin tX \rangle$ and $\langle \cos tX \rangle$. The Lévy inversion formula, see Feller[20], implies that $p(dx)$ is determined by $\phi(it)$. Thus, it is not even necessary to know the expectation on all of $\mathcal{C}(\mathbf{R})$ to determine the measure $p(dx)$.

1.3 INDEPENDENCE AND CONVOLUTION

The pair (X, Y) of random variables corresponds to a measure $p(dx, dy)$ on $\mathbf{R}^2$. Denote the individual distributions on $\mathbf{R}$, the *marginal distributions* of X and Y, by p_1 and p_2 respectively. Then X and Y are *independent* if

$$p(dx, dy) = p_1(dx) p_2(dy)$$

This means, if $\psi(x, y) \in \mathcal{C}(\mathbf{R}^2)$ is of the form $\psi(x, y) = f(x)g(y)$, $f, g \in \mathcal{C}(\mathbf{R})$, then

$$
\begin{aligned}
\langle \psi(X, Y) \rangle &= \int_{-\infty}^{\infty} \int_{-\infty}^{\infty} \psi(x, y) \, p(dx, dy) \\
&= \int_{-\infty}^{\infty} \int_{-\infty}^{\infty} f(x)g(y) \, p_1(dx) p_2(dy) \\
&= \langle f(X) \rangle \cdot \langle g(Y) \rangle
\end{aligned}
$$

that is, $\langle f(X)g(Y) \rangle = \langle f(X) \rangle \langle g(Y) \rangle$ — the expectation of the product is the product of the expectations. In general:

1.3.1 Definition. The family $(X_1, \ldots, X_N)$ of random variables is *independent* if their (joint) distribution on $\mathbf{R}^N$ factors as the product of their individual marginal distributions on $\mathbf{R}$, i.e.,

$$p(dx_1, dx_2, \ldots, dx_N) = p_1(dx_1)p_2(dx_2) \cdots p_N(dx_N)$$

where, for $1 \le j \le N$, X_j has distribution p_j.

Essentially equivalent to this definition, we have, in terms of expectations,

1.3.2 Proposition. Let $(X_1, \ldots, X_N)$ be independent. Let $\psi(x_1, x_2, \ldots, x_N) \in \mathcal{C}(\mathbf{R}^N)$ be of the form $f_1(x_1)f_2(x_2) \cdots f_N(x_N)$, $f_1, f_2, \ldots, f_N \in \mathcal{C}(\mathbf{R})$, then

$$
\begin{aligned}
\langle \psi(X_1, \ldots, X_N) \rangle &= \langle f_1(X_1)f_2(X_2) \cdots f_N(X_N) \rangle \\
&= \langle f_1(X_1) \rangle \cdot \langle f_2(X_2) \rangle \cdots \langle f_N(X_N) \rangle
\end{aligned}
$$

We are particularly interested when this works for polynomials as well. Fubini's theorem implies that if the marginal distributions $p_1, p_2, \ldots, p_N \in \mathcal{P}^*$, then the Proposition above holds with the replacement of the functions f_j by polynomials. Furthermore, we have

1.3.3 Proposition. Let $(X_1, X_2, \ldots, X_N)$ be a family of independent random variables each having distribution in $\mathcal{P}^*$. Let $S = X_1 + \cdots + X_N$ denote their sum. Then the distribution of S is in $\mathcal{P}^*$. The moment generating function of S is the product of the moment generating functions of the X_j. I.e.,

$$\langle e^{zS} \rangle = \langle e^{zX_1} \rangle \cdots \langle e^{zX_N} \rangle$$

Proof: Follows via Fubini's theorem, writing

$$e^{zS} = e^{zX_1} \cdot e^{zX_2} \cdot \ldots \cdot e^{zX_N}$$

∎

Consider X, Y independent, with distributions p_1, p_2 respectively. Then, with $S = X + Y$,

$$\langle f(S) \rangle = \langle f(X+Y) \rangle = \int_{-\infty}^{\infty} \int_{-\infty}^{\infty} f(x+y)\, p_1(dx) p_2(dy) = \int_{-\infty}^{\infty} f(s)\, p(ds)$$

where $p(ds)$ is the distribution of their sum. We have

1.3.4 Definition. Let p_1 and p_2 be probability measures on **R**. Their *convolution* denoted $p_1 * p_2$ is the distribution given by

$$\int_{-\infty}^{\infty} f(x)\, (p_1 * p_2)(dx) = \int_{-\infty}^{\infty} \int_{-\infty}^{\infty} f(y+y')\, p_1(dy) p_2(dy')$$

for $f \in \mathcal{C}(\mathbf{R})$.

This extends to families $(X_1, \ldots, X_N)$.

1.3.5 Proposition. Let $(X_1, \ldots, X_N)$ be independent with corresponding distributions $p_1, \ldots, p_N$. Then the distribution of their sum $S_N = X_1 + \cdots + X_N$ is given by the convolution $p_1 * \cdots * p_N$.

Proof: We have only to remark that the operation $*$ is associative, since it corresponds directly to addition of the random variables. ∎

For densities, we have:

1.3.6 Proposition. Let $p_1(dx) = p_1(x)\, dx$ and $p_2(dx) = p_2(x)\, dx$ be given by densities. Then the convolution $(p_1 * p_2)(dx)$ has density

$$(p_1 * p_2)(x) = \int_{-\infty}^{\infty} p_1(x-y) p_2(y)\, dy$$

Proof: With

$$\int_{-\infty}^{\infty} f(x)\, (p_1 * p_2)(dx) = \int_{-\infty}^{\infty} \int_{-\infty}^{\infty} f(y'+y'')\, p_1(y') p_2(y'')\, dy'\, dy''$$

changing variables: $x = y' + y''$, $y = y''$ yields the result. ∎

Remark. The operation $*$ is commutative and associative. It can be defined for arbitrary (absolutely) integrable functions on **R**. Thus, if f and g are integrable on **R**, their convolution is defined as

$$(f * g)(x) = \int_{-\infty}^{\infty} f(x-y) g(y)\, dy$$

1.4 CONVOLUTION SEMIGROUPS

Starting with a probability measure $p(dx)$, set $p_1 = p$ and define inductively the convolution powers

$$p_{N+1} = p * p_N$$

The family of measures $\{p_N\}_{N \geq 0}$ is the *convolution semigroup* generated by p. If $p \in \mathcal{P}^*$, with $\exp(H(z)) = \int \exp(zx) p(dx)$, then Prop. 1.3.3 yields

$$\int_{-\infty}^{\infty} e^{zx} \, p_N(dx) = e^{NH(z)}$$

When this can be extended to a continuous-parameter family is the subject of *infinitely divisible laws*, which we will not present here; see Feller[20]. However, we can say that *if* $\{\exp(tH(z))\}_{t \geq 0}$ corresponds to a family of probability measures $p_t(dx)$, with

$$e^{tH(z)} = \int_{-\infty}^{\infty} e^{zx} \, p_t(dx)$$

then it determines a one-parameter convolution semigroup of measures p_t, satisfying $p_{t+s} = p_t * p_s$, for all $s, t \geq 0$, with p_0 defined as the delta function at 0.

In general discussion, we use t as the parameter. It is to be understood that only for the infinitely-divisible laws mentioned above does this give probability measures for all real $t \geq 0$. When discrete time is considered explicitly, we denote the parameter by N, as above.

Note that $p \in \mathcal{P}^*$ implies that elements of the convolution semigroup generated by p all lie in $\mathcal{P}^*$, since it is just a matter of the analyticity of $H(z)$ at 0. We have, for $p_t \in \mathcal{P}^*$, the moments $\mu_n(t)$ given via the expansion

$$e^{tH(z)} = \sum_{n=0}^{\infty} \frac{z^n}{n!} \mu_n(t) \tag{1.4.1}$$

1.4.1 Random walks

If X is a random variable with distribution $p(dx)$, then corresponding to the convolution powers of p are sums of independent random variables all having the same distribution, p. That is, let X_j, $j \geq 1$, denote a sequence of independent, identically distributed random variables, with distribution p. Then, from Prop. 1.3.5, the sequence of sums $S_N = X_1 + \cdots + X_N$ have corresponding distributions p_N.

This family of random variables is a stochastic process called the *random walk* corresponding to p. The interpretation is as follows. Start at the origin on $\mathbf{R}$.

At each unit of time, jump an amount determined at random according to the distribution p. Then S_N gives your location after N units of time have elapsed.

If the process can be run continuously, i.e., for real $t \geq 0$, it is called a (time-homogeneous) *process with independent increments* . It may still move by jumps, but it can also move in a continuous fashion. In Chapter 5, we will meet specific examples of these processes, in both discrete and continuous time. They are special types of *Markov processes*. (References: Breiman [10], Feller[20], Kac[28], Karlin and Taylor[29], Ross[42], Revuz[41] (Markov chains in particular))

II.   Moment systems

Moment systems are families of polynomials corresponding to convolution semigroups of probability measures. We extend the term to families of polynomials having a similar analytic and algebraic structure. In this section all of these systems have a natural structure in terms of representations of the HW algebra.

2.1   MOMENT POLYNOMIALS

Given a convolution semigroup of measures in $\mathcal{P}^*$, they define an *evolution* or *flow* acting on functions as in Ch. 3, Def. 2.3.2.6. With $\exp(tH(z)) = \int \exp(zx)\, p_t(dx)$, we have

$$e^{tH(D)} f(x) = \int_{-\infty}^{\infty} f(x+y)\, p_t(dy)$$

for $f \in \mathcal{C}(\mathbf{R}) \cup \mathcal{P}$. For $f(x) = x^n$, we have:

2.1.1 Definition.   The *moment polynomials* corresponding to the convolution family p_t are given by

$$h_n(x,t) = \int_{-\infty}^{\infty} (x+y)^n\, p_t(dy)$$

Remark. Of course, one could define moment polynomials for a single p, without considering the corresponding flow.

Recalling eq. (1.4.1),

2.1.2 Proposition.   *The moment polynomials are given in terms of the moments* $\mu_k(t)$:

$$h_n(x,t) = \sum_{k=0}^{n} \binom{n}{k} x^{n-k} \mu_k(t)$$

Proof: Expand $(x + y)^n$ by the binomial theorem and integrate. ∎

The action $e^{tH} x^n$ can be determined in various ways.

1. Since $H(0) = 0$, $H(D)$ acts nilpotently on $\mathcal{P}$, see Ch. 1, §1.3. Thus,

$$e^{tH(D)} x^n = \sum_{k=0}^{\infty} \frac{t^k H(D)^k}{k!} x^n$$

is well-defined as a finite series for each n.

2. We can expand, eq. (1.4.1), $\exp(tH(z)) = \sum (z^k/k!)\,\mu_k(t)$. Thus

$$h_n(x,t) = e^{tH(D)} x^n = \sum_{k=0}^{\infty} \frac{\mu_k(t)}{k!} D^k x^n = \sum_{k=0}^{n} \binom{n}{k} x^{n-k} \mu_k(t)$$

in agreement with Prop. 2.1.2. Cf., calculation of matrix elements, Ch. 1, Prop. 2.2.3.1.

From point #1., we see that

2.1.3 Proposition. *For all $t \in \mathbf{R}$, the moment polynomials exist and the action $e^{tH} x^n$ is well-defined.*

This is regardless of the existence of corresponding probability measures p_t for non-integer t. The definition of $\mu_k(t)$ extends accordingly to $t \in \mathbf{R}$.

We compute the generating function.

2.1.4 Proposition. *The generating function for the moment polynomials*

$$G_t(a,x) = e^{ax+tH(a)} = \sum_{n=0}^{\infty} \frac{a^n}{n!} h_n(x,t)$$

exists for all a in some neighborhood of the origin.

Proof: The series, eq. (1.4.1), converges absolutely for z in a neighborhood of 0. Thus, via Prop. 2.1.2,

$$\sum_{n=0}^{\infty} \frac{a^n}{n!} h_n(x,t) = \sum_{n=0}^{\infty} \frac{a^n}{n!} \sum_{k=0}^{n} \binom{n}{k} x^{n-k} \mu_k(t) = \left(\sum_{n=0}^{\infty} \frac{a^n}{n!} \right) \left(\sum_{k=0}^{\infty} \frac{a^k \mu_k(t)}{k!} \right)$$

and the result follows. ∎

Related to this is a feature of the operator calculus for D, cf. Ch. 1, §§1.3,2.2. Since $\exp(ax)$ are eigenfunctions of D, we have $\phi(D)\exp(ax) = \phi(a)\exp(ax)$ for all a in the domain of ϕ. For polynomial ϕ, this holds for all a. We check that this is consistent with Ch. 3, Def. 2.3.2.6, used above as well.

2.1.5 Proposition. For $\phi(z) = \int e^{zx} p(dx)$, $\phi(D)e^{ax} = \phi(a)e^{ax}$ holds.

Proof: By definition, Ch. 3, 2.3.2.6,

$$\phi(D)e^{ax} = \int_{-\infty}^{\infty} e^{a(x+y)} p(dy) = e^{ax} \int_{-\infty}^{\infty} e^{ay} p(dy) = e^{ax} \phi(a)$$

by definition of ϕ. ∎

This provides a useful method for verifying operator identities. Namely, by applying to functions e^{ax}, we reduce to analytic relations. Another way to view this is to think of e^{ax} as the generating function for the powers x^n, so an operator relation that holds on functions e^{ax}, that is analytic in a, yields results for the x^n. This is the point of view alluded to in Ch. 1, §1.3. E.g., here we have

$$e^{tH(D)} e^{ax} = e^{ax+tH(a)} = G_t(a, x)$$

giving the generating function for the moment polynomials $e^{tH(D)} x^n$.

We conclude this section with some observations.

2.1.6 Proposition. For $n \geq 0$, the moment polynomial $h_n(x, t)$ is the solution $u(x, t)$ to the evolution equation

$$\frac{\partial u}{\partial t} = H(D)u$$

with initial function $u(x, 0) = x^n$.

Proof: This follows from observation #1 above. ∎

This gives the interpretation of the h_n as evolved powers under the flow generated by H. Similarly, note that $G_t(a, x)$ satisfies $\partial u/\partial t = Hu$, with $u(x, 0) = e^{ax}$.

2.1.7 Proposition. The moment polynomials are a basis for $\mathcal{P}$. In fact, the h_n are monic, with $\deg(h_n) = n$, for $n \geq 0$.

Proof: Since $\mu_0(t) = 1$, the leading term of $h_n(x, t)$ is x^n. ∎

Now we introduce the representation of the HW algebra on $\mathcal{P}$ corresponding to the basis $\psi_n = h_n(x, t)$. The vacuum state Ω is the constant polynomial $1 = h_0(x, t)$. The raising operator R_t satisfies $R_t h_n(x, t) = h_{n+1}(x, t)$. We will discuss R_t in detail in §2.3. The lowering operator, V, satisfying $V h_n(x, t) = n h_{n-1}(x, t)$ is easily found.

2.1.8 Proposition.   *The lowering operator for the standard HW representation acting on the moment polynomials h_n is given by D, independent of t.*

Proof: We have $h_n = \exp(tH(D))x^n$. Since D and $H(D)$ commute,
$$Dh_n(x,t) = De^{tH(D)} x^n = e^{tH(D)} Dx^n = e^{tH(D)} nx^{n-1}$$
and the result follows.   ∎

Now we look more closely at the algebraic and analytic features not depending on the probabilistic structure.

2.2 APPELL SYSTEMS

First,

2.2.1 Definition.   *Appell polynomials*   denote a sequence $h_n(x)$ of elements in $\mathcal{P}$ with the following properties:
  1. For $n \geq 0$, $\deg(h_n(x)) = n$.

  2. With $D = d/dx$, $Dh_n(x) = nh_{n-1}(x)$.

That is, these are polynomial sequences, bases for $\mathcal{P}$, having lowering operator $V = D$. In particular, the import of Props. 2.1.7, 2.1.8 is that moment polynomials are Appell polynomials.

Given the constant $c_0 = h_0$, we can use property #2 and, successively introducing constants of integration, inductively build up the polynomials.

2.2.2 Proposition.   *Appell polynomials are determined by a sequence $\{c_n\}_{n \geq 0}$ with $c_0 \neq 0$. They are given by*
$$h_n(x) = \sum_{k=0}^{n} \binom{n}{k} c_k x^{n-k}$$

The condition $c_0 \neq 0$, corresponds to #1 of the definition. In general, the theory can be developed from the point of view of formal power series. Here we take the analytic approach. So, we assume that the sequence $\{c_n\}$ is such that $\phi(z) = \sum c_n z^n/n!$ converges in a neighborhood of the origin ($z \in \mathbf{C}$). Then, as in Prop. 2.1.4, we have the generating function
$$G(z,x) = \sum_{n=0}^{\infty} \frac{z^n}{n!} h_n(x) = e^{zx} \phi(z)$$

Since $\phi(0) = c_0 \neq 0$, we can represent $\phi(z)$ near 0 in the form $c_0 e^{H(z)}$, as for moment polynomials.

Remark. We will take the normalization $c_0 = 1$ throughout.

We can generate a flow, using $\exp(tH(D))$. Thus, we call *Appell systems*   the associated polynomial sequences $\{h_n(x,t)\}$. Similarly, when the $\{c_n\}$ are a sequence of moments of a probability measure, we have   *moment systems* .

2.2.3 Theorem. *Given any function $H(z)$, analytic at 0, we have a corresponding Appell system $\{h_n(x,t)\}_{n \geq 0}$ defined for all $x, t \in \mathbf{R}$. The polynomials are given by the generating function*

$$G_t(a, x) = e^{ax + tH(a)}$$

and by the evolution generated by $H(D)$:

$$h_n(x, t) = e^{tH(D)} x^n$$

Note that, in fact, we could take x, t to be complex variables.

Now to clarify the HW structure.

2.3 GENERALIZED MOMENT SYSTEMS. HW REPRESENTATIONS. RAISING OPERATOR

We start with the definition of GMS — generalized moment systems.

2.3.1 Definition. Given $H(z)$, analytic at 0, with $H(0) = 0$, a *generalized moment system (GMS)* with generator $H(z)$ is a (one-parameter) family of polynomial sequences evolving under the group generated by the operator $H(D)$.

One can also consider generators H involving x as well as D (e.g., see §4.4.2).

Remark. By convention, H alone, as in e^{tH}, indicates the operator $H(D)$, while for the function $H(z)$, e.g., we show the argument explicitly.

To specify the GMS we need initial conditions *time-zero polynomials* . Then these evolve forwards or backwards in time according to the flow e^{tH} .

At present, we consider a GMS with generator $H(z)$ and time-zero polynomials $\{x^n\}_{n \geq 0}$. At time t, we have the basis $\{h_n(x,t)\}_{n \geq 0}$, given by $h_n(x,t) = e^{tH} x^n$. We want a HW representation, i.e., operators V_t, R_t such that $R_t h_n = h_{n+1}$ and $V_t h_n = n h_{n-1}$. Proposition 2.1.8 gives $V_t = D$, for all t. To find R_t, form the generating function:

$$G_t(a, x) = \sum_{n=0}^{\infty} \frac{a^n}{n!} h_n(x, t) = \sum_{n=0}^{\infty} \frac{a^n}{n!} (R_t)^n 1 = e^{a R_t} 1$$

Then we find R_t by differentiating with respect to a. I.e., treating (x, t) as parameters, $u(a) = G_t(a, x)$ satisfies

$$\frac{\partial u}{\partial a} = R_t u, \qquad u(0) = 1$$

This leads to the result:

2.3.2 Proposition.   *The raising operator R_t is given by*

$$R_t = x + tH'(D)$$

Proof: From Prop. 2.1.4, $G_t(a, x) = \exp(ax + tH(a))$. Thus,

$$R_t G_t = \frac{\partial G_t}{\partial a} = (x + tH'(a))e^{ax + tH(a)}$$
$$= (x + tH'(D))\, G_t$$

replacing a by D acting on e^{ax}.   ∎

2.3.3 Corollary.   *The polynomials $h_n(x, t)$ are given recursively, via $h_{n+1} = R_t h_n$, as $h_n(x, t) = (x + tH'(D))^n 1$.*

One can use the expansion of $H'(z)$ to calculate $H'(D)x^n$, and hence the action of $H'(D)$ on polynomials. Cf. below, §2.3.2 and §IV.

2.3.1 Heisenberg flow and HW representations

To see the HW structure, we take another approach.

2.3.1.1 Proposition.   *The raising operator R_t has the form*

$$R_t = e^{tH}\, x e^{-tH}$$

Proof: Since e^{tH} is a group, we can invert the relation $h_n = e^{tH}\, x^n$, writing $x^n = e^{-tH}\, h_n$. Then

$$R_t h_n = h_{n+1} = e^{tH}\, x^{n+1} = e^{tH}\, x(x^n)$$
$$= e^{tH}\, x(e^{-tH}\, h_n) = (e^{tH}\, x e^{-tH})h_n$$

∎

The HW commutation rules, Ch. 1, Prop. 2.1.1, give

$$R_t = e^{tH}\, x e^{-tH} = (x e^{tH} + tH' e^{tH})e^{-tH} = x + tH'$$

as in Prop. 2.3.2. And we see directly, via $(R_t)^n = e^{tH}\, x^n e^{-tH}$ that

$$(R_t)^n 1 = e^{tH}\, x^n e^{-tH} 1 = e^{tH}\, x^n$$

since $H(0) = 0$, so that $H(D)1 = 0$. And since D commutes with $H(D)$, we have the lowering operator, $V_t = D = e^{tH}\, D e^{-tH}$.

Remark. The action on an algebra by conjugation: $\xi \to \xi(t) = e^{tH} \xi e^{-tH}$ is an *automorphism*. It preserves the algebraic structure in one-one correspondence. E.g.,

$$\xi_1(t)\xi_2(t) = e^{tH} \xi_1 e^{-tH} e^{tH} \xi_2 e^{-tH} = e^{tH} \xi_1\xi_2 e^{-tH} = (\xi_1\xi_2)(t)$$

Thus, commutation relations are preserved as well:

$$[\xi_1, \xi_2] = \xi_3 \Leftrightarrow [\xi_1(t), \xi_2(t)] = \xi_3(t)$$

From this approach, we see the following construction of a GMS.

1. Start with an HW representation, operators $\{R_0, V_0\}$, on the time-zero space with basis $\{\psi_n\}_{n \geq 0}$.

2. Evolve the operators according to the automorphism generated by H:

$$R_t = e^{tH} R_0 e^{-tH}, \qquad V_t = e^{tH} V_0 e^{-tH}$$

   so that $[V_t, R_t] = 1$, $\forall t$.

3. At time t, the basis is given by $\{(R_t)^n \psi_0\}_{n \geq 0}$.

This corresponds to the *Heisenberg picture* in quantum mechanics. We will thus call the automorphism group, conjugation by e^{tH}, acting on an algebra, the *Heisenberg flow* generated by H.

2.3.2 Raising operator

We will develop a 'calculus' of the raising operator $R_t = e^{tH} x e^{-tH}$. To compute directly $(R_t)^n = (x + tH'(D))^n$, we use the rule (cf. Ch. 1, Prop. 2.1.1, and §IV below)

$$[\phi(D), x] = \phi'(D)$$

repeatedly: $[H(D), x] = H'(D)$, $[H'(D), x] = H''(D)$, etc. E.g.,

$$(R_t)^2 = (x + tH'(D))(x + tH'(D))$$
$$= x^2 + 2xtH'(D) + tH''(D) + t^2 H'(D)^2$$

The problem is to find a formula for $(R_t)^n$ as an operator.

In general, $R_t = e^{tH} x e^{-tH}$ implies $\phi(R_t) = e^{tH} \phi(x) e^{-tH}$, which holds immediately for polynomials. We first calculate the group generated by R_t, $\exp(aR_t)$, as operators, not just acting on 1.

2.3.2.1 Proposition. *The exponential of R_t is given by*

$$e^{aR_t} = e^{ax} e^{t[H(D+a) - H(D)]}$$

Proof: We use the idea of the discussion following Prop. 2.1.5. I.e., we find $e^{aR_t} e^{zx}$ and then replace z by D. We have

$$e^{aR_t} e^{zx} = e^{tH(D)} e^{ax} e^{-tH(D)} e^{zx}$$
$$= e^{tH(D)} e^{ax} e^{-tH(z)} e^{zx}$$
$$= e^{tH(z+a)-tH(z)} e^{ax} e^{zx}$$

and, first pulling through e^{ax}, the result follows by replacing z by D as remarked above. ∎

Now we proceed to find $(R_t)^n$. First we have

2.3.2.2 Definition. The *moment operators* $\eta_k(D,t)$ are given by the functions $\eta_k(z,t)$

$$\eta_k(z,t) = e^{-tH(z)} \left(\frac{\partial}{\partial z}\right)^k e^{tH(z)}$$

The terminology comes from the expansion $\exp(tH(z)) = \sum(z^k/k!)\mu_k(t)$, so that

$$\mu_k(t) = \eta_k(0,t) \tag{2.3.2.1}$$

(see also Problem 2, §7.1, of Ch. 6.)

2.3.2.3 Theorem. *The operator* $(R_t)^n$ *is given in terms of the moment operators by:*

$$(R_t)^n = (x + tH'(D))^n = \sum_{k=0}^{n} \binom{n}{k} x^{n-k} \eta_k(D,t)$$

Proof: Expand the result of Prop. 2.3.2.1:

$$e^{aR_t} e^{zx} = e^{ax} e^{t[H(z+a)-H(z)]} e^{zx}$$

in powers of a, computing $(\partial/\partial a)^n|_{a=0}$. Using Leibniz' rule, $(\partial/\partial a)^n(f(a)g(a)) = \sum \binom{n}{k} f^{(k)}(a) g^{(n-k)}(a)$, we have

$$(R_t)^n e^{zx} = \sum_{k=0}^{n} \binom{n}{k} x^{n-k} e^{-tH(z)} \left(\frac{\partial}{\partial a}\right)^k \bigg|_{a=0} e^{tH(z+a)} e^{zx}$$

Comparing with Def. 2.3.2.2, the result follows. ∎

Applying this result to $h_0 = 1$, using eq. (2.3.2.1), recovers the form of the polynomials $h_n(x,t)$ directly.

2.4 PROPERTIES OF GMS

We summarize the features of the GMS considered to this point.

1. The generator $H(z)$ is a given function analytic in a neighborhood of the origin (in $\mathbf{C}$).

2. The expansion $\exp(tH(z)) = \sum(z^n/n!)\mu_n(t)$ determines the generalized moments $\mu_n(t)$.

3. The time-zero polynomials are $\{x^n\}_{n\geq 0}$.

4. The initial raising and lowering operators are $R_0 = x$, $V_0 = D$. At time t, the raising and lowering operators are

$$R_t = e^{tH} R_0 e^{-tH} = e^{tH} x e^{-tH} = x + tH'(D)$$
$$V_t = e^{tH} V_0 e^{-tH} = D$$

For given t, $R_t h_n = h_{n+1}$, $V_t h_n = n h_{n-1}$, giving a representation of the HW algebra.

5. At time t, the polynomials

$$h_n(x,t) = (R_t)^n 1 = \sum_{k=0}^{n} \binom{n}{k} x^{n-k}\mu_k(t)$$

The generating function $G_t(a,x) = \sum(a^n/n!)h_n(x,t)$ is

$$G_t(a,x) = e^{ax+tH(a)} = e^{aR_t} 1$$

6. As operators, we have: the exponential of R_t

$$e^{aR_t} = e^{tH} e^{ax} e^{-tH} = e^{ax} e^{t[H(D+a)-H(D)]}$$

and the powers of R_t

$$(R_t)^n = \sum_{k=0}^{n} \binom{n}{k} x^{n-k}\eta_k(D,t)$$

where the moment operators η_k are given by the functions

$$\eta_k(z,t) = e^{-tH(z)} \left(\frac{\partial}{\partial z}\right)^k e^{tH(z)}$$

with the generalized moments $\mu_n(t) = \eta_n(0,t)$.

7. The functions $h_n(x,t)$ are solutions of the evolution equation $\partial u/\partial t = Hu$. I.e., $h_n(x,t)$ is the solution to

$$\frac{\partial u}{\partial t} = H(D)u, \qquad u(x,0) = x^n$$

The generating function $G_t(a,x)$ satisfies

$$\frac{\partial u}{\partial t} = H(D)u, \qquad u(x,0) = e^{ax}$$

2.4.1 Addition formulas

The generating functions, $\exp(tH(z)) = \sum(z^n/n!)\mu_n(t)$ and $\exp(ax + tH(a)) = \sum(a^n/n!)h_n(x,t)$, give addition formulas generalizing the binomial theorem — which corresponds to the exponential relation $e^a e^b = e^{a+b}$. The following relations are thus often referred to as *identities of binomial type*. (This approach is featured in works of Rota, e.g., Rota[43].)

2.4.1.1 Proposition. *The generalized moments $\{\mu_n(t)\}$ and the associated polynomials $\{h_n(x,t)\}$ satisfy the addition formulas:*

$$\mu_n(t+s) = \sum_{k=0}^{n} \binom{n}{k} \mu_{n-k}(t)\,\mu_k(s)$$

$$h_n(x+y,t+s) = \sum_{k=0}^{n} \binom{n}{k} h_{n-k}(x,t)\,h_k(x,s)$$

Proof: For the first line, use

$$e^{tH(z)}\, e^{sH(z)} = e^{(t+s)H(z)}$$

and expand. Similarly for the second line.   ∎

These correspond to representations of the translation group on $\mathbf{R}^1$ and $\mathbf{R}^2$ respectively. Cf. matrix elements, Ch. 1, §2.2.3.

III.   Radial moment systems

Radial moment systems refer to GMS, generalized moment systems, based on the sl(2) algebra. The initial algebra has standard basis $\{\Delta, R, \rho\}$. Given the generator $H(z)$, the flow is generated by the operator $H(\Delta)$.

Remark. Unless otherwise indicated, we use the realization of Chapter 3, §2.4. So that, for $R^n\Omega$ we have the basis x^n, with Δ acting as $x\,(d^2/dx^2) + c\,(d/dx)$.

We expect the Hankel-Mellin transform, $\int \mathcal{I}_c(zx)\,p(dx)$, associated with the Bessel function, to play a rôle analogous to and as well as the Fourier-Laplace transform, the moment generating function, $\int \exp(zx)\,p(dx)$, associated with the exponential function. We have the scheme

Operator	Eigenfunctions	Transform
D	e^{zx}	$\int e^{zx}\,p(dx)$
Δ	$\mathcal{I}_c(zx)$	$\int \mathcal{I}_c(zx)\,p(dx)$

In Chapter 3, we had the Fock space structure and the interpretation of these eigenfunctions as reproducing kernels.

We begin with a basic proposition which shows that the class $\mathcal{P}^*$ of measures, Def. 1.1.3, is suitable for the radial moment systems.

3.1 Lemma.   If $p \in \mathcal{P}^*$, then the Hankel-Mellin transform exists and is analytic in a neighborhood of the origin in the complex plane.

Proof: Let the moment generating function of p be analytic for $|z| < \varepsilon$. Let $0 < t < \varepsilon$. Then the following integrals are all finite:

$$\int_{-\infty}^{\infty} e^{ty}\, p(dy) = \int_0^{\infty} e^{ty}\, p(dy) + \int_{-\infty}^0 e^{ty}\, p(dy)$$

$$= \int_0^{\infty} e^{ty}\, p(dy) + \int_0^{\infty} e^{-ty}\, p(-dy)$$

Call these last two integrals #1 and #2. Similarly,

$$\int_{-\infty}^{\infty} e^{-ty}\, p(dy) = \int_0^{\infty} e^{-ty}\, p(dy) + \int_0^{\infty} e^{ty}\, p(-dy)$$

these integrals being #3 and #4 respectively. Then

$$\int_{-\infty}^{\infty} e^{t|y|}\, p(dy) = (\#1) + (\#4) < \infty$$

Thus, for all $-\varepsilon < t < \varepsilon$, $\int \exp(t|y|)\, p(dy) < \infty$. Now, we have the estimate, Ch. 3, Prop. 2.4.1.2, #3: $\mathcal{I}_c(r) < \exp(r/c)$, for $r > 0$. Thus,

$$\int_{-\infty}^{\infty} |\mathcal{I}_c(zy)|\, p(dy) \leq \int_{-\infty}^{\infty} e^{|zy|/c}\, p(dy) < \infty$$

for $|z| < c\varepsilon$. To see analyticity, writing $\mathcal{I}_c(zy)$ as a power series, the same estimate gives uniform absolute convergence for $|z| < c\varepsilon$.   ∎

Thus, for the remainder of this section, we assume all measures appearing to be in $\mathcal{P}^*$.

3.1   RADIAL CONVOLUTIONS

There are several ways of thinking about convolution of measures. Here are two features to consider:

1. The convolution $p_1 * p_2$ is characterized by the fact that it is multiplicative for moment generating functions:

$$\phi_j(z) = \int_{-\infty}^{\infty} e^{zx}\, p_j(dx), \quad j = 1, 2 \Rightarrow \phi_1(z)\phi_2(z) = \int_{-\infty}^{\infty} e^{zx}\,(p_1 * p_2)(dx)$$

2. It is determined by the action of the translation operator $\exp(yD)$ on functions $f \in \mathcal{C}(\mathbf{R})$. Integrate $e^{yD} f(x)$ with respect to $p_1(dx)p_2(dy)$:

$$\int_{-\infty}^{\infty}\int_{-\infty}^{\infty} e^{yD} f(x)\, p_1(dx)p_2(dy) = \int_{-\infty}^{\infty}\int_{-\infty}^{\infty} f(x+y)\, p_1(dx)p_2(dy)$$

$$= \int_{-\infty}^{\infty} f(s)\,(p_1 * p_2)(ds)$$

We define *radial convolution* to have these features for the corresponding Hankel-Mellin transforms (#1) and for the action $\mathcal{I}_c(y\Delta)f(x)$ (#2).

3.1.1 Definition. The *radial translation* given as $\mathcal{I}_c(y\Delta)f(x)$ is denoted $f(x \oplus y)$.

3.1.2 Definition. The *radial convolution* of p_1 and p_2, denoted $p_1 \circ p_2$ is defined by the relation

$$\int_{-\infty}^{\infty}\int_{-\infty}^{\infty} f(x \oplus y)\, p_1(dx)p_2(dy) = \int_{-\infty}^{\infty} f(r)\,(p_1 \circ p_2)(dr)$$

for all $f \in \mathcal{C}(\mathbf{R})$.

Cf. Prop. 3.1.4 below and remark following. For $p_1, p_2 \in \mathcal{P}^*$, this holds for all polynomials as well.

We can write this in terms of random variables:

$$\int_{-\infty}^{\infty}\int_{-\infty}^{\infty} f(x \oplus y)\, p_1(dx)p_2(dy) = \langle f(X \oplus Y)\rangle$$

Recalling Ch. 3, Lemma 2.4.2.4,

3.1.3 Proposition. *The binomial theorem for radial translation:*

$$(x \oplus y)^n = \sum_{k=0}^{n} \binom{n}{k} \frac{(c)_n}{(c)_{n-k}(c)_k} x^{n-k} y^k$$

Note that this corresponds to the multiplication formula

$$\mathcal{I}_c(x \oplus y) = \mathcal{I}_c(x)\,\mathcal{I}_c(y)$$

which is clear from series expansions (cf. the proof below).

3.1.4 Proposition.   *Properties of radial convolution:*

1. Let $\phi_1(z) = \int \mathcal{I}_c(zx)\, p_1(dx)$, $\phi_2(z) = \int \mathcal{I}_c(zx)\, p_2(dx)$, then

$$\int_{-\infty}^{\infty} \mathcal{I}_c(zx)\,(p_1 \circ p_2)(dx) = \phi_1(z)\phi_2(z)$$

   i.e., the Hankel-Mellin transforms are multiplicative.

2. Let $h(x)$ be a polynomial. Then, for $c > \frac{1}{2}$,

$$\langle h(X \oplus Y)\rangle = \int_{-\infty}^{\infty}\int_{-\infty}^{\infty}\int_{0}^{2\pi} h(x + y - 2\sqrt{xy}\cos\vartheta)\,(\sin^2\vartheta)^{c-1}\,\frac{d\vartheta\, p_1(dx)p_2(dy)}{2B(c-1/2,1/2)}$$

   Proof: For #1, use the fact that $\mathcal{I}_c(zx)$ is an eigenfunction of Δ, $\Delta\mathcal{I}_c(zx) = z\mathcal{I}_c(zx)$. Thus,

$$\mathcal{I}_c(z(x \oplus y)) = \mathcal{I}_c(y\Delta)\mathcal{I}_c(zx) = \mathcal{I}_c(zy)\mathcal{I}_c(zx)$$

and #1 follows. For #2, the result follows from the integral formula, Ch. 3, Theorem 2.4.2.6.   ∎

Remark. Of course, we can extend #2 to more general functions. Note that the operation $\oplus$ is commutative and associative. It is an example of *generalized translation*. (See, e.g., Levitan[35].)

3.1.1 Radial convolution semigroups

   As for ordinary convolution, we take $p = p_1$ and build the radial convolution semigroup: $p_N^{\circ} = p \circ p \circ \cdots \circ p$. Writing $\exp(H(z)) = \int \mathcal{I}_c(zx)\, p(dx)$, we have, by the multiplicative property,

$$e^{tH(z)} = \int_{-\infty}^{\infty} \mathcal{I}_c(zx)\, p_t^{\circ}(dx)$$

where, as usual, we write t for the (time) parameter. We have the semigroup property: $p_{t+s} = p_t \circ p_s$. Denoting the moments of p_t° by $\mu_n(t)$, we have

3.1.1.1 Proposition.   *For radial convolution we have the moment expansion*

$$e^{tH(z)} = \sum_{n=0}^{\infty} \frac{z^n}{n!\,(c)_n}\,\mu_n(t)$$

Proof: Expanding $\mathcal{I}_c(zx)$:

$$\int_{-\infty}^{\infty} \mathcal{I}_c(zx)\, p_t^\circ(dx) = \sum_{n=0}^{\infty} \frac{z^n}{n!\,(c)_n} \int_{-\infty}^{\infty} x^n p_t^\circ(dx)$$

and the result follows.   ■

This gives, analogously to the formula in Prop. 2.4.1.1:

3.1.1.2 Proposition.   *The moments $\mu_n(t)$ satisfy the addition formula*

$$\mu_n(t+s) = \sum_{k=0}^{n} \binom{n}{k} \frac{(c)_n}{(c)_{n-k}\,(c)_k}\, \mu_{n-k}(t)\, \mu_k(s)$$

3.2   sl(2) MOMENT SYSTEMS

We consider first the general structure of sl(2) generalized moment systems, sl(2) GMS, then we look at some realizations. Start with $H(z)$, analytic at 0, $H(0) = 0$, as generator, with the corresponding operator $H(\Delta)$. The initial operators $\{\Delta_0, R_0, \rho_0\}$ are just $\{\Delta, R, \rho\}$, a standard sl(2) basis.

The Heisenberg flow is

$$R_t = e^{tH} R e^{-tH}\,, \qquad \rho_t = e^{tH} \rho e^{-tH}\,, \qquad \Delta_t = e^{tH} \Delta e^{-tH}$$

We find

3.2.1 Proposition.   *With $H = H(\Delta)$, the operators evolve via the Heisenberg flow according to:*

$$R_t = R + t\,\rho H' + (tH'' + t^2 H'^2)\Delta$$
$$\rho_t = \rho + 2t\,\Delta H'$$
$$\Delta_t = \Delta$$

Proof: Use the sl(2) commutation rules, Ch. 1, Prop. 3.2.1, extended to analytic functions.   ■

The point is that for every t fixed, $\{\Delta_t, R_t, \rho_t\}$ is a standard sl(2) basis. At time zero, we take the standard representation with basis $\{x^n\}_{n\geq 0}$:

$$R x^n = x^{n+1}, \qquad \rho\, x^n = (c+2n)x^n, \qquad \Delta x^n = n(c+n-1)x^{n-1}$$

The sl(2) GMS polynomials are then given by

$$h_n(x,t) = e^{tH} x^n, \qquad \text{with } H = H(\Delta)$$

3.2.2 Proposition.   *Properties of sl(2) GMS:*

1. *At time t, the basis is given by $h_n(x,t) = (R_t)^n 1$.*

2. *For each t, we have the sl(2) representation:*

$$R_t h_n = h_{n+1}, \qquad \rho_t h_n = (c + 2n)h_n, \qquad \Delta_t h_n = n(c + n - 1)h_{n-1}$$

3. *The polynomial $h_n(x,t)$ satisfies the evolution equation*

$$\frac{\partial u}{\partial t} = H(\Delta)u, \qquad u(x,0) = x^n$$

Note that various realizations of sl(2) can be used. For example, we will consider in §3.3 the realization of Δ as half the Laplacian acting on radial functions.

As we have seen, from the probabilistic point of view, to a given measure $p(dx)$ one can consider either the ordinary moment generating function or else the Hankel-Mellin transform. That is, to a given $p(dx)$ we have *two* possible functions $H(z)$ to consider. These correspond to *exponential systems* and to *Bessel systems*, respectively, which we discuss in the following subsections. It is important to clarify that from the point of view of GMS, the sl(2) structure of these systems is the same. It is the choice of generator $H(z)$ that effectively makes the difference. From the probabilistic point of view, it is the choice of transform, hence the convolution structure, which is the distinguishing feature.

3.2.1 Exponential systems

These correspond to ordinary convolution, $p_{t+s} = p_t * p_s$, with $e^{tH(z)} = \int e^{zy} p(dy)$, and the operator formalism

$$e^{tH(\Delta)} = \int_{-\infty}^{\infty} e^{y\Delta} p_t(dy)$$

Here, for general $H(z)$, the usual expansion $\exp(tH(z)) = \sum (z^n/n!)\mu_n(t)$ gives the correspondence between $H(z)$ and the generalized moments. Now the matrix elements of the group generated by Δ, Ch. 1, Prop. 3.3.2.1, come into play. We have

3.2.1.1 Theorem.   *The GMS polynomials $h_n(x,t)$ for the exponential system are given by*

$$h_n(x,t) = \sum_{k=0}^{n} \binom{n}{k} \frac{(c)_n}{(c)_{n-k}} x^{n-k} \mu_k(t)$$

Proof: In the probabilistic case, the matrix elements, Ch. 1, 3.3.2.1, give

$$e^{y\Delta} x^n = \sum_{k=0}^{n} \binom{n}{k} \binom{n+c-1}{n-k}(n-k)!\, y^{n-k} x^k = \sum_{k=0}^{n} \binom{n}{k} \frac{(c)_n}{(c)_{n-k}} x^{n-k} y^k$$

and integration with respect to $p_t(dy)$ yields the result. In general, for $h_n(x,t) = e^{tH(\Delta)} x^n$, the expansion of $\exp(tH(z))$, replacing z by Δ, and the action

$$\Delta^k x^n = n^{(k)}(c+n-1)^{(k)} x^{n-k} = \frac{n!}{(n-k)!} \frac{(c)_n}{(c)_{n-k}} x^{n-k}$$

gives the result. (See Ch. 1, Prop. 3.3.1.1, and the proof of Prop. 3.3.2.1.)   ∎

Referring to the list of polynomials at the end of the Introduction, we make

3.2.1.2 Definition.   The *Laguerre moment polynomials* are given by

$$l_n(x,c) = \sum_{k=0}^{n} \binom{n}{k} \frac{(c)_n}{(c)_k} x^k$$

Remark. With $L_n(x,c)$ the Laguerre polynomials in the reference list, we have $L_n(x,c) = (-1)^n l_n(-x,c)$. (We are using c here instead of t, since t has a different use in the present context.)

This gives the formulation

3.2.1.3 Theorem.   *In the probabilistic case, the moment polynomials $h_n(x,t)$ for the exponential system have the form*

$$h_n(x,t) = \int_{-\infty}^{\infty} l_n(x/y,c)\, y^n p_t(dy)$$

Proof: From Def. 3.2.1.2:

$$y^n l_n(x/y,c) = \sum_{k=0}^{n} \binom{n}{k} \frac{(c)_n}{(c)_k} x^k y^{n-k}$$

Now substitute $k \to n-k$ and integrate with respect to p_t. By Theorem 3.2.1.1, the result follows.   ∎

Thus, for the sl(2) exponential moment systems the Laguerre polynomials play the role of the binomial powers $(x-1)^n$. Implicit in the above formulation is the following

3.2.1.4 Proposition.   *The Laguerre moment polynomials satisfy*

$$y^n l_n(x/y,c) = e^{y\Delta} x^n$$

while the Laguerre polynomials satisfy

$$y^n L_n(x/y,c) = e^{-y\Delta} x^n$$

with $\Delta = x\,(d^2/dx^2) + c\,(d/dx)$.

3.2.2 Bessel systems

These correspond to radial convolution, $p_{t+s} = p_t \circ p_s$, with $e^{tH(z)} = \int \mathcal{I}_c(zy) \, p_t^\circ(dy)$, and the operator formalism

$$e^{tH(\Delta)} = \int_{-\infty}^{\infty} \mathcal{I}_c(y\Delta) \, p_t^\circ(dy)$$

Here, for general $H(z)$, the generalized moments $\mu_n(t)$ are determined by the expansion, Prop. 3.1.1.1,

$$e^{tH(z)} = \sum_{n=0}^{\infty} \frac{z^n}{n! \, (c)_n} \mu_n(t) \tag{3.2.2.1}$$

3.2.2.1 Theorem. *The GMS polynomials $h_n(x,t)$ for the Bessel system are given by*

$$h_n(x,t) = \sum_{k=0}^{n} \binom{n}{k} \frac{(c)_n}{(c)_{n-k} \, (c)_k} \, x^{n-k} \mu_k(t)$$

Proof: For the probabilistic case, Ch. 3, Lemma 2.4.2.4 gives

$$\mathcal{I}_c(y\Delta) x^n = \sum_{k=0}^{n} \binom{n}{k} \frac{(c)_n}{(c)_{n-k} \, (c)_k} \, x^{n-k} y^k$$

and hence the result by integration with respect to $p_t(dy)$. For general $H(z)$, this follows from $h_n(x,t) = e^{tH(\Delta)} x^n$ via eq. (3.2.2.1), replacing z by Δ. ∎

This suggests

3.2.2.2 Definition. The *Hankel moment polynomials*

$$\Phi_n(x,c) = \sum_{k=0}^{n} \binom{n}{k} \frac{(c)_n}{(c)_{n-k} \, (c)_k} \, x^k$$

Remark. These have the expression in terms of hypergeometric functions:

$$\Phi_n(x,c) = {}_2F_1 \left(\begin{matrix} -n, 1-c-n \\ c \end{matrix} \,\middle|\, x \right)$$

which follows readily using eq. (3.2) of the Introduction.

We have, via Theorem 3.2.2.1,

3.2.2.3 Theorem. *In the probabilistic case, the moment polynomials $h_n(x,t)$ for the Bessel system have the form*

$$h_n(x,t) = \int_{-\infty}^{\infty} \Phi_n(x/y,c) \, y^n p_t^\circ(dy)$$

3.3 HEAT EQUATION. HERMITE POLYNOMIALS. LAGUERRE POLYNOMIALS

The *heat equation* is the evolution equation with generator half the Laplacian. I.e., on $\mathbf{R}^N$,

$$\frac{\partial u}{\partial t} = \tfrac{1}{2} \sum_{j=1}^{N} \frac{\partial^2 u}{\partial x_j^2}$$

with $u(\mathbf{x}, 0)$ to be specified. For $N = 1$, we have

$$\frac{\partial u}{\partial t} = \frac{1}{2} \frac{\partial^2 u}{\partial x^2}$$

the evolution with generator $H(z) = z^2/2$. We consider the corresponding moment system. This is called the *Hermite moment system* since the moment polynomials are closely related to Hermite polynomials as will be seen shortly.

3.3.1 Proposition.   *For the Hermite moment system:*

1. *The moments $\mu_n(t)$ are zero for n odd, and*

$$\mu_{2k}(t) = (2t)^k (\tfrac{1}{2})_k$$

2. *The Hermite moment polynomials   are*

$$h_n(x,t) = \sum_{k} \binom{n}{2k} x^{n-2k} (2t)^k (\tfrac{1}{2})_k$$

3. *The raising operator is $R_t = x + tD$.*

4. *The generating function is*

$$e^{ax + a^2 t/2} = \sum_{n=0}^{\infty} \frac{a^n}{n!} h_n(x,t)$$

Proof: Expand $e^{tH(z)}$, $H(z) = z^2/2$:

$$e^{tz^2/2} = \sum_{k=0}^{\infty} \frac{z^{2k} t^k}{2^k k!} = \sum_{n=0}^{\infty} \frac{z^n}{n!} \mu_n(t)$$

by definition. This gives $\mu_{2k}(t) = t^k (2k)!/(2^k k!)$, yielding the stated result, via Intro., eq. (3.1). The rest follows from the general theory of moment systems, §2.4.

∎

3.3.2 Definition. The Hermite polynomials $H_n(x,t)$ are given by

$$H_n(x,t) = h_n(x,-t)$$

where $h_n(x,t)$ are the Hermite moment polynomials.

Remark. Thus, Hermite polynomials are the *time-reversed* moment polynomials. They satisfy the equation $\partial u/\partial t + (1/2)\,\partial^2 u/\partial x^2 = 0$.

For the heat equation on $\mathbf{R}^N$, with the notation $D_j = \partial/\partial x_j$, the generator is $\frac{1}{2}\sum D_j^2$.

3.3.3 Proposition. With $H = \frac{1}{2}\sum D_j^2$, on $\mathbf{R}^N$, we have

$$e^{tH}\, x_1^{n_1} x_2^{n_2} \cdots x_N^{n_N} = h_{n_1}(x_1,t)h_{n_2}(x_2,t)\cdots h_{n_N}(x_N,t)$$

where $h_n(x,t)$ are the Hermite moment polynomials.

Proof: This follows from the observation that the operators D_j commute and that D_j affects only functions of x_j. ■

This gives the solution to the heat equation with initial function a given monomial $x_1^{n_1} x_2^{n_2} \cdots x_N^{n_N}$.

The Laguerre moment polynomials, Prop. 3.2.1.4, satisfy

$$t^n l_n(x/t,c) = e^{t\Delta}\, x^n$$

with $\Delta = xD^2 + cD$, the sl(2) lowering operator. This realization comes from the sl(2) calculus, Ch. 1, §3.3.1, given by $\Delta\Omega = 0$, $\rho\Omega = c\Omega$, $R^n\Omega = x^n$. Before proceeding further, we state some properties of the l_n.

3.3.4 Proposition. Properties of $l_n(x,c)$:

1. In terms of hypergeometric functions:

$$l_n(x,c) = (c)_n\, {}_1F_1\left(\begin{array}{c}-n\\c\end{array}\,\Big|\,-x\right)$$

2. We have the generating function for l_n:

$$\sum_{n=0}^{\infty} \frac{a^n}{n!}\, l_n(x,c) = (1-a)^{-c}\exp\left(\frac{a}{1-a}\,x\right)$$

Proof: #1 follows from rewriting the definition, Def. 3.2.1.2. We prove #2 as an application of the sl(2) exponential commutation rule, Ch. 1, Prop. 3.3.2:

$$e^{t\Delta} e^{aR} = \exp(\frac{a}{1-at} R)(1-at)^{-\rho} \exp(\frac{t}{1-at} \Delta)$$

Applying this to Ω, with $t = 1$, and $\Delta\Omega = 0$, $\rho\Omega = c\Omega$ as usual, yields:

$$e^{\Delta} e^{aR} \Omega = (1-a)^{-c} \exp(\frac{a}{1-a} R)\Omega$$

With $e^{\Delta} R^n \Omega = e^{\Delta} x^n = l_n(x, c)$, the result follows. ∎

For any realization of sl(2) acting on Ω as noted above, sl(2) calculus gives the relation

$$e^{t\Delta} R^n \Omega = t^n l_n(R/t, c)\Omega$$

Now we will use the realization with Δ as half the Laplacian on $\mathbf{R}^N$, Ch. 1, eq. (3.1.2). I.e.,

$$\Delta = \tfrac{1}{2}\sum_{j=1}^{N} D_j^2, \qquad R = \tfrac{1}{2}\sum_{j=1}^{N} x_j^2, \qquad \rho = \sum_{j=1}^{N} x_j \frac{\partial}{\partial x_j} + \frac{N}{2}$$

Note that $c = N/2$ here. We will write $r^2 = \sum x_j^2$. Thus, we have explicitly:

3.3.5 Proposition. *The Laguerre moment polynomials, with generator Δ on $\mathbf{R}^N$, satisfy*

$$e^{t\sum D_j^2/2} r^{2n} = (2t)^n l_n(r^2/2t, N/2)$$

with $r^2 = \sum x_j^2$.

That is, the Laguerre moment polynomials give the solution to the heat equation on $\mathbf{R}^N$ with initial functions r^{2n}.

By using both the HW and sl(2) calculus, we can find some interesting connections between the Hermite moment polynomials and Laguerre moment polynomials.

Remark. It is easy to check that the Hermite moment polynomials satisfy the scaling property

$$h_n(x, t) = t^{n/2} h_n(x/\sqrt{t}, 1)$$

Thus, for the remainder of this discussion we write $h_n(x)$ for $h_n(x, 1)$.

3.3.6 Proposition. *The even-order Hermite moment polynomials satisfy*

$$h_{2n}(x) = 2^n l_n(x^2/2, 1/2)$$

Proof: Starting from HW operators, write

$$h_{2n}(x) = e^{D^2/2} x^{2n} = e^{\Delta} (2(x^2/2))^n = 2^n e^{\Delta} (x^2/2)^n$$

in terms of sl(2) operators, which gives the result, cf. Prop. 3.3.5.   ∎

Similarly, using the realization on $\mathbf{R}^N$.

3.3.7 Proposition.   *The Laguerre moment polynomials have the expansion in terms of even-order Hermite moment polynomials:*

$$2^n l_n(r^2/2, N/2) = \sum_{k_1 + \cdots + k_N = n} \binom{n}{k_1, \ldots, k_N} h_{2k_1}(x_1) \cdots h_{2k_N}(x_N)$$

where $r^2 = \sum x_j^2$.

Proof: Start with the left-hand side:

$$2^n e^{\Delta} (r^2/2)^n = e^{\frac{1}{2} \sum D_j^2} (\sum_{j=1}^{N} x_j^2)^n$$

and the result follows from Prop. 3.3.3 after expanding via the multinomial theorem.
∎

We can also find a result for odd-order polynomials h_{2n+1} in one dimension.

3.3.8 Proposition.   *Odd-order Hermite moment polynomials satisfy*

$$h_{2n+1}(x) = 2^n x\, l_n(x^2/2, 3/2)$$

Proof: The HW calculus gives

$$h_{2n+1}(x) = e^{D^2/2} x^{2n+1} = 2^{n+1/2} e^{D^2/2} (x^2/2)^{n+1/2}$$

From sl(2) calculus,

$$e^{\Delta} R^{n+1/2}\Omega = \sum_k \binom{n+1/2}{k} \frac{(c)_{n+1/2}}{(c)_{n-k+1/2}} R^{n-k+1/2}\Omega$$

In one dimension, $c = 1/2$, so

$$e^{D^2/2} (x^2/2)^{n+1/2} = \sum_k \binom{n+1/2}{k} \frac{(1/2)_{n+1/2}}{(1/2)_{n-k+1/2}} (x^2/2)^{n-k+1/2}$$

Observe that

$$\frac{(1/2)_{n+1/2}}{(1/2)_{n-k+1/2}} = \frac{\Gamma(n+1)}{\Gamma(n-k+1)} = \frac{n!}{(n-k)!}$$

so that the sum is indeed finite, with $0 \le k \le n$. Now, substitute $k \to n-k$. Noting

$$(n+1/2)^{(n-k)} = (n+1/2)\cdots(k+3/2) = \frac{(3/2)_n}{(3/2)_k}$$

gives

$$\sum_{k=0}^{n} \binom{n}{k} \frac{(3/2)_n}{(3/2)_k} (x^2/2)^{k+1/2}$$

and, with the factor $2^{n+1/2}$, the result follows. ∎

We thus have, combining Props. 3.3.4, 3.3.6, 3.3.8:

3.3.9 Proposition.   *Hermite moment polynomials can be expressed in terms of hypergeometric functions:*

$$h_{2n}(x) = 2^n (\tfrac{1}{2})_n \, {}_1F_1\left(\begin{matrix} -n \\ 1/2 \end{matrix} \,\bigg|\, -\frac{x^2}{2}\right)$$

$$x^{-1} h_{2n+1}(x) = 2^n (\tfrac{3}{2})_n \, {}_1F_1\left(\begin{matrix} -n \\ 3/2 \end{matrix} \,\bigg|\, -\frac{x^2}{2}\right)$$

To conclude, we combine Props. 3.3.7 and 3.3.8 to find the identity

3.3.10 Proposition.   *The Hermite moment polynomials satisfy*

$$(x^2+y^2+z^2)^{-1/2} h_{2n+1}(\sqrt{x^2+y^2+z^2}) = \sum_{j+k+l=n} \binom{n}{j,k,l} h_{2j}(x)h_{2k}(y)h_{2l}(z)$$

Proof: The result follows from Prop. 3.3.8 via the choice $N = 3$ in Prop. 3.3.7. ∎

IV.   Holomorphic canonical variables

This gives some 'geometric flavor' to the theory. Namely, how changing variables comes into play. We will work locally, i.e., with functions holomorphic in a neighborhood of 0 in the complex plane.

Remark. We will use letters x, e.g., as well as z to denote complex variables.

4.1 Definition. Denote by $\mathcal{H}$ the set of functions holomorphic in some neighborhood (in general, depending on the function) of $0 \in \mathbf{C}$.

We define an action of the functions $\phi \in \mathcal{H}$ on polynomials in x. We denote the action by . First let z act as differentiation:

$$z \cdot x^n = nx^{n-1}$$

and, with the natural action of multiplication by x, $x \cdot x^n = x^{n+1}$, we have the HW relations $[z, x] = 1$. Now to extend to arbitrary ϕ.

4.2 Definition. The action of $\phi \in \mathcal{H}$ is defined by

$$\phi(z) \cdot x^n = \sum_{k=0}^{n} \binom{n}{k} x^{n-k} \phi^{(k)}(0)$$

where $\phi^{(k)}$ denotes the kth derivative of ϕ.

This definition is, of course, based on the HW commutation rules, Ch. 1, Prop. 2.2.2. In terms of the operator $D = d/dx$, we have $\phi(z) \cdot x^n = \phi(D)x^n$. And for given ϕ, the action extends to exponential functions so that $\phi(z) \cdot e^{ax} = \phi(a)e^{ax}$ holds for all a in the domain of ϕ. In particular,

4.3 Proposition. For $\phi \in \mathcal{H}$, for the action on polynomials,

$$[\phi(z), x] = \phi'(z)$$

Proof: We know this follows for any polynomial ϕ, from the properties of the HW algebra. And acting on a polynomial of degree n, only derivatives of ϕ up to order n are involved. So we may replace ϕ by a corresponding polynomial. Alternatively, one can check the relation directly from Def. 4.2, via Pascal's triangle relations for the binomial coefficients. ∎

Remark. Note that if $\phi \in \mathcal{H}$ satisfies $\phi(0) \neq 0$, then $1/\phi \in \mathcal{H}$ so that acting on polynomials, $\phi(z)$ is 1-1 with inverse $1/\phi(z)$.

4.1 V OPERATOR

We want to find a holomorphic realization, i.e., involving elements of $\mathcal{H}$, of the operators V, R giving a representation of the HW algebra on a space of polynomials. The principal feature is:

the expression $e^{ax} 1$ is invariant under holomorphic changes of coordinates

(Note that when acting on 1, we will drop the dot notation, since ordinarily the 1 would not be written.) Let us clarify what this means. Let $V \in \mathcal{H}$ satisfy $V(0) = 0$, $V'(0) \neq 0$, i.e., V is locally invertible (in the sense of composition of functions) in a neighborhood of 0. Then, if we change variables $z \to V(z)$, in terms of the *dual* variable, ξ, we have the equality:

$$e^{V(a)\xi} 1 = e^{ax} 1 \qquad (4.1.1)$$

How is ξ determined? We want the HW relations $[V, \zeta] = 1$ to hold. I.e., ξ is a holomorphic realization of the raising operator R. Let $W(z) = 1/V'(z) \in \mathcal{H}$. Since $[V(z), x] = V'(z)$, we see immediately

4.1.1 Proposition.   Given $V(z) \in \mathcal{H}$ satisfying $V(0) = 0$, $V'(0) \neq 0$, let
$\xi = xW(z)$. Then we have

$$[V, \xi] = 1$$

i.e., V and ξ are a boson pair.

4.1.2 Definition.   Let $V \in \mathcal{H}$ satisfy $V(0) = 0$, $V'(0) \neq 0$. Let $\xi = x(1/V'(z))$.
Then V, ξ are *canonical variables* and ξ is the *canonical dual* of V.

   Remark. There are many operators, η, say, such that $[V, \eta] = 1$, e.g., let $\eta = \xi + \phi(z)$ for any $\phi \in \mathcal{H}$, cf., the raising operator $R_t = x + tH'$ of HW moment systems.

Remark. We use the standard notations:

$$W(z) = 1/V'(z), \qquad \xi = xW(z), \qquad U(V(z)) = z, \; V(U(v)) = v$$

i.e., $U(v)$ denotes the *inverse function* of V. Of course, $U \in \mathcal{H}$.

   Now observe that $\xi^n 1$ provides us with a basis for the space of polynomials.

4.1.3 Proposition.   For $n \geq 0$, $\xi^n 1$ is a polynomial of degree n in x. The
leading coefficient is $W(0)^n$.

   Proof: Starting with $\xi^0 1 = 1$, $\xi 1 = xW(0)$, we have inductively:

$$\xi(a_n x^n + r_n(x)) = x a_n \sum_{k=0}^{n} \binom{n}{k} x^{n-k} W^{(k)}(0) + \tilde{r}_{n+1}(x)$$

$$= a_n x^{n+1} W(0) + r_{n+1}(x)$$

where $a_n \neq 0$ is the leading coefficient of $\xi^n 1$ and $\deg r_n < n$. Thus, $\deg \tilde{r}_{n+1} \leq 1 + \deg r_n < n + 1$, and the result follows.   ■

Thus,

4.1.4 Proposition.   With the basis $\psi_n = \xi^n 1$, we have the HW representation

$$\xi \psi_n = \psi_{n+1}, \qquad V\psi_n = n\psi_{n-1}$$

with the vacuum $\psi_0 = 1$.

4.2 EXPONENTIAL FORMULA

Now we give the basic result, clarifying eq. (4.1.1).

4.2.1 Theorem.   *The exponential formula for ξ:*

$$e^{v\xi}\, 1 = e^{xU(v)}$$

Proof: $u = e^{v\xi}\, 1$ satisfies

$$\frac{\partial u}{\partial v} = \xi u, \qquad u(0) = 1$$

Note that if $x = 0$, then u is identically 1 as a function of v. Via HW calculus, we have $V(z) \cdot u = vu$, and thus, $z \cdot u = U(v)u$. I.e., as a function of x, u satisfies $Du = U(v)u$, $u|_0 = 1$. Hence the result.   ∎

We have the extension

4.2.2 Corollary.   *The action of $e^{t\xi}$ on e^{ax} is given by*

$$e^{t\xi} \cdot e^{ax} = e^{xU(t+V(a))}$$

Proof: Writing $e^{ax} = \exp(V(a)\xi)1$, apply $e^{t\xi}$:

$$e^{t\xi} \cdot e^{ax} = e^{t\xi}\, e^{V(a)\xi}\, 1$$
$$= e^{(t+V(a))\xi}\, 1 = e^{xU(t+V(a))}$$

by Theorem 4.2.1.   ∎

Put into words: from the point of view of the HW calculus based on the operators V, ξ, on the exponential $\exp(v\xi)1$, V acts as multiplication by v. From the point of view of the calculus based on D and x, the operator $V(D)$ acts on $\exp(ax)$ by multiplication by $V(a)$. So if $a = U(v)$, with $U(v)$ inverse to $V(z)$, the operator $V(D)$ acts as multiplication by v.

In the next section we will look at this from the point of view of expansions, i.e., generating functions.

4.3 EXPANSION FORMULA

First some notation

4.3.1 Definition. Given canonical variables V, ξ as functions of (z, x), denote the basic polynomials by $\xi_n(x) = \xi^n 1$.

Now we will present a direct way of seeing how the *canonical HW calculus* works.

4.3.2 Theorem. *Let V, ξ be canonical variables. Then we have the expansion:*

$$e^{ax} = \sum_{n=0}^{\infty} \frac{V(a)^n}{n!} \xi_n(x)$$

Proof: The question is that of identifying the coefficients $\xi_n(x)$ in the expansion. Differentiate with respect to a:

$$xe^{ax} = \sum_{n=0}^{\infty} \frac{nV(a)^{n-1}V'(a)}{n!} \xi_n(x)$$

With $W(a) = 1/V'(a)$, readjusting the summation index:

$$xW(a)e^{ax} = \sum_{n=0}^{\infty} \frac{V(a)^n}{n!} \xi_{n+1}(x)$$

Replacing a by D yields the recursion

$$\xi_{n+1}(x) = xW(D)\xi_n(x)$$

which, with $\xi_0(x) = 1$, says that $\xi_n(x) = \xi^n 1$ as required. ∎

We see that the substitution $a = U(v)$ recovers the formulation of Theorem 4.2.1.

Remark. For the remainder of this Chapter, we will return to the convention of Chapter 1, where a 'general function' f is any polynomial or a sum of exponential functions with polynomial coefficients. Extensions to various classes of analytic functions can be done, e.g., via constructions such as Fock spaces, or by finding analytic conditions that give convergence of series, say, in the polynomial basis $\xi_n(x)$. (References: Boas&Buck[8], De Branges[12], De Branges&Rovnyak[13].)

4.3.3 Theorem. *In the canonical variables V, ξ we have the generalized Taylor formula*

$$f(x + y) = \sum_{n=0}^{\infty} \frac{\xi_n(y)}{n!} V(D)^n f(x)$$

Proof: Write the expansion in Theorem 4.3.2 with the substitutions $x \to y$, $a \to D$ and apply to $f(x)$:

$$f(x + y) = e^{yD} f(x) = \sum_{n=0}^{\infty} \frac{\xi_n(y)}{n!} V(D)^n f(x)$$

as required. ∎

In the next section we discuss various techniques and applications of this calculus.

4.4 TECHNIQUES AND APPLICATIONS

First we discuss some general techniques related to the expansion formula, Theorem 4.3.3.

To calculate $V(D)^n$, write $V(z) = ze^{H(z)}$, taking the normalization $V'(0) = 1$ here. Then we have immediately, via the theory of moment systems:

4.4.1 Proposition.   Let $V(z) = ze^{H(z)}$, where $e^{H(z)}$ is the Fourier-Laplace transform of a measure $p \in \mathcal{P}^*$. Then

$$V(D)^n f(x) = \int_{-\infty}^{\infty} f^{(n)}(x+y)\, p_n(dy)$$

where p_n denotes the convolution semigroup generated by p.

Another formulation is useful for calculating $\xi_n(x)$.

4.4.2 Proposition.   Let $V(z) = ze^{-H(z)}$, for some $H \in \mathcal{H}$. Then

$$\xi = x(1 - zH'(z))^{-1} e^{H(z)}$$

Proof: Calculate $W(z) = 1/V'(z)$.   ∎

Here the idea is that we can expand in geometric series:

$$(1 - zH'(z))^{-1} = \sum_{n=0}^{\infty} z^n H'(z)^n$$

The following subsections provide some further illustrations of the canonical HW calculus.

4.4.1 Lagrange inversion

The question here is: given a function $\phi \in \mathcal{H}$, with non-vanishing derivative at the origin, find the power series expansion of the inverse function $\phi^{-1}(z)$, and, further, the expansion of the composition $f(\phi^{-1}(z))$ for arbitrary $f \in \mathcal{H}$.

Taking V to be the function in question, the problem is to find series expansions for functions of the form $f(U(v))$, $f \in \mathcal{H}$. The HW calculus applies nicely.

4.4.1.1 Theorem.   Let V, ξ be canonical variables. Then, for $f \in \mathcal{H}$,

$$f(U(v)) = \sum_{n=0}^{\infty} \frac{v^n}{n!}\, \xi_n(D) f(0)$$

Proof: Write Theorem 4.2.1 in the form, with $x \to a$,

$$e^{aU(v)} = \sum_{n=0}^{\infty} \frac{v^n}{n!} \xi_n(a)$$

Now put $a \to D$, and apply to $f(x)$ to get

$$e^{U(v)D} f(x) = f(x + U(v)) = \sum_{n=0}^{\infty} \frac{v^n}{n!} \xi_n(D)f(x)$$

and evaluate at $x = 0$. ∎

Remark. The proof holds for arbitrary $f \in \mathcal{H}$: since $\deg \xi_n(x) = n$, at each step, only a finite number of derivatives of f are involved in calculating successively the coefficients of the expansion.

Remark. In general, we refer to §V for exercises and examples to be worked out. In this and the next section, we will indicate two examples that should help the reader to see how the techniques described work. Some more examples referring to these sections are in §V.

Example. (This example is well-known in graph theory and combinatorics see, e.g., Comtet[11]) Take $V(z) = ze^{-z}$. So $\xi = x(1-z)^{-1}e^z$. Then $\xi_n(x) = x(x+n)^{n-1}$, $n \geq 1$, with $\xi_0(x) = 1$. The easiest way to verify this is via the observation that ξ maps the set of polynomials onto the set of polynomials of the form $xf(x)$ in 1–1 fashion. Thus, one verifies

$$\xi_{n-1}(x) = \xi^{-1} \cdot \xi_n(x) = e^{-z}(1-z)x^{-1} \cdot \xi_n(x)$$

Theorem 4.2.1 reads

$$e^{xU(v)} = 1 + x \sum_{n=1}^{\infty} \frac{v^n}{n!} (x+n)^{n-1} \qquad (4.4.1.1)$$

which yields the interesting formulas:

$$e^{ax} = 1 + x \sum_{n=1}^{\infty} \frac{a^n e^{-na}}{n!} (x+n)^{n-1}$$

$$U(v) = \sum_{n=1}^{\infty} \frac{v^n}{n!} n^{n-1}$$

the first via Theorem 4.3.2, and the second follows most readily from eq. (4.4.1.1) via $U(v) = \lim_{x \to 0} (e^{xU(v)} - 1)/x$.

4.4.2 Exponentiation of $x\psi(z)$

The canonical HW calculus is directly concerned with the case where $\xi = xW(z)$, with $W(0) \neq 0$. An interesting situation arises when we want to calculate, say, $\eta^n \cdot f(x)$, with $\eta = x\psi(z)$, where $\psi(0) = 0$.

Let $Y(z) = \int dz/\psi(z)$ as an indefinite integral. Now, suppose that $V(z) = e^{Y(z)}$ is a canonical coordinate, i.e.,

$$\lim_{z \to 0} Y'(z)e^{Y(z)} \neq 0$$

Since we require $V(0) = 0$, $Y = \log V$ is necessarily singular at the origin. Now, $\psi(z) = 1/Y'(z) = V(z)W(z)$, so that

$$\eta = xV(z)W(z) = \xi V(z)$$

On the basis $\xi_n(x) = \xi^n 1$ we have

$$\eta \cdot \xi_n(x) = \xi V(z) \cdot \xi_n(x) = n\,\xi_n(x)$$

That is, η is the *number operator* for the representation (Ch. 1, §2.4).

4.4.2.1 Theorem. *Let $\eta = x\psi(z)$. Let $Y(z) = \int dz/\psi(z)$ satisfy*

$$\lim_{z \to 0} e^{Y(z)} = 0, \qquad \lim_{z \to 0} Y'(z)e^{Y(z)} \neq 0$$

Then, taking as canonical variables

$$V(z) = e^{Y(z)}, \qquad \xi = x(1/Y'(z))e^{-Y(z)}$$

we have

$$e^{t\eta} \cdot f(x) = \sum_{n=0}^{\infty} \frac{\xi_n(x)}{n!}\, e^{nt}\, e^{nY(D)} f(0)$$

Proof: By Theorem 4.3.3, substituting $y \to x$, $x \to 0$,

$$f(x) = \sum_{n=0}^{\infty} \frac{\xi_n(x)}{n!}\, e^{nY(D)} f(0)$$

Since η acts as the number operator, the result follows. ∎

Example. Let $\eta = x(e^z - 1)$. Then $Y(z) = \log(1 - e^{-z})$, $V(z) = 1 - e^{-z}$. And we find

$$\xi_n(x) = (xe^z)^n 1 = x(x+1)\cdots(x+n-1) = (x)_n$$

Thus, we have the expansion

$$e^{t\eta} f(x) = \sum_{n=0}^{\infty} \frac{(x)_n}{n!}\, e^{nt}\, (1 - e^{-D})^n f(0)$$

4.4.3 Generalized moments and canonical variables

In the theory of GMS, we use the expansion

$$e^{tH(z)} = \sum_{n=0}^{\infty} \frac{z^n}{n!} \mu_n(t)$$

for $H \in \mathcal{H}$ to define the generalized moments $\mu_n(t)$. Comparing this with Theorem 4.2.1, we see that $\mu_n(t)$ are the canonical basis $\xi_n(x)$ via the correspondence

$$x \leftrightarrow t, \qquad U(v) \leftrightarrow H(z)$$

That is, H is the inverse of the canonical coordinate V. The underlying variables are now (t, T), with the action

$$T \cdot f(t) = \frac{d}{dt} f(t)$$

This leads to

4.4.3.1 Proposition.   Let $H \in \mathcal{H}$, $H(0) = 0$, $H'(0) \neq 0$. Let $V(T) = H^{-1}(T)$. Then, $\xi = t\, H'(V(T))$ is the canonical dual, and the generalized moments are given by $\xi_n(t)$, i.e.,

$$\mu_n(t) = \xi^n 1$$

Proof: Since $V(T) = H^{-1}(T)$, $W(T) = H'(H^{-1}(T)) = H'(V(T))$. The result follows by the canonical HW calculus.   ∎

This nicely ties in with the theory of moment systems. We can read this in reverse, to get a representation of the basis $\xi_n(x)$.

4.4.3.2 Theorem.   Let V, ξ be canonical variables such that $e^{U(v)}$ is the Fourier-Laplace transform of $p \in \mathcal{P}^*$. Then, for all x for which the convolution semigroup p_x is defined,

$$\xi_n(x) = \int_{-\infty}^{\infty} y^n\, p_x(dy)$$

4.4.4 Addition formulas

As with the generalized moments, Prop. 2.4.1.1, we have

4.4.4.1 Proposition.   The polynomials $\xi_n(x)$ satisfy the identity of binomial type

$$\xi_n(x + y) = \sum_{k=0}^{n} \binom{n}{k} \xi_{n-k}(x)\, \xi_k(y)$$

Notice that this is a relation in the tensor product of two copies of the HW representation, cf. Ch. 3, §III. That is, we consider commuting pairs of operators $(x, \partial/\partial x)$, $(y, \partial/\partial y)$. Then we have the canonical operator pairs:

$$(V(\partial/\partial x), \xi(x)), \qquad (V(\partial/\partial y), \xi(y))$$

with $\xi(x) = xW(\partial/\partial x)$ and likewise for $\xi(y)$. Similarly, we put, with $s = x + y$, $\xi(s) = sW(\partial/\partial s)$. An interesting formulation of the addition formula is the following.

4.4.4.2 Theorem. *Let* $\xi(x+y)f(x+y) = \xi(s)f(s)$ *evaluated at* $s = x+y$. *Then*

$$\phi(\xi(x) + \xi(y))f(x + y) = \phi(\xi(x + y))f(x + y)$$

Proof: We check for $\phi(x) = e^{tx}$, $f(s) = e^{as}$. Thus, Corollary 4.2.2 applied to the variables x and y individually:

$$e^{t(\xi(x)+\xi(y))} e^{a(x+y)} = e^{(x+y) U(t+V(a))}$$
$$= e^{sU(t+V(a))}$$
$$= e^{t\xi(s)} e^{as}$$

∎

In this formulation, Prop. 4.4.4.1 results by taking $\phi(x) = x^n$, $f(x) = 1$.

Remark. Some texts dealing primarily with special functions: Abramowitz&Stegun[1], Lebedev[32], Nikiforov&Uvarov[37], Rainville[40], Watson[47]. For stochastic processes on Lie groups, see Heyer[14]. Also see Feinsilver&Schott[17], [18], and, especially, [19]. The book Boas&Buck[8] discusses expansions in various systems of polynomials many of which are (generalized) moment systems, or closely related to them. Chapter 4 and the next two chapters are based on Feinsilver[15].

V.   Exercises and examples

5.1 EXERCISES

1. Prove Proposition 1.2.2

2. Show, via Proposition 1.3.3, that if X_j, $1 \le j \le N$, are independent and have the same distribution with moment generating function $\phi(z)$, then $S_N = X_1 + X_2 + \cdots + X_N$ has moment generating function $\phi(z)^N$.

3. Let $p(x) = e^{-x}$ for $x \in [0, \infty)$, the exponential density. Find the N-fold convolution $p * p * \cdots * p$.

4. Show directly that convolution (as defined in the remark following Proposition 1.3.6) is commutative and associative.

5. Show that from equation (1.4.1), the expansion of $\exp(tH(z))$, follows

$$\mu_1(t) = t\mu, \qquad \sigma^2(t) = t\sigma^2$$

where $\mu = \mu_1(1)$, $\sigma^2 = \sigma^2(1)$, and $\sigma^2(t) = \mu_2(t) - \mu_1(t)^2$, as usual.

6. Use points #1 and #2 following Proposition 2.1.2 to find the moment polynomials $h_n(x, t)$ for the following generators $H(z)$: i) z ii) $z^2/2$ iii) $z^3/3$ iv) z^p/p, integer $p > 0$. Write out $h_0, h_1, \ldots, h_5$ explicitly.

7. Verify that your polynomials in Problem 6 satisfy Proposition 2.1.6, the appropriate evolution equation.

8. a. As in §2.2, verify that your polynomials (Problem 6) satisfy $Dh_n = nh_{n-1}$.

   b. Show directly from the formula given in Proposition 2.2.2 that $Dh_n = nh_{n-1}$.

9. Referring to §2.3, calculate the raising operators for the generators in Problem 6. In each case, use the raising operator to calculate $h_0, \ldots, h_5$ recursively. E.g., for $R_t = x + tD$, $h_0 = 1$, $h_1 = (x + tD)1 = x$, etc.

10. Discuss the remark following Proposition 2.3.1.1 concerning automorphisms.

11. In the context of §2.3.2, calculate R_t^3 directly as done there for R_t^2.

12. Use Proposition 2.3.2.1 to find $\exp(aR_t)$ for the generators in Problem 6. And as well for $H(z) = \log \cosh z$, $H(z) = -\log(1 - z)$.

13. Calculate the moment operators $\eta_k(z, t)$, $1 \le k \le 4$, for the generators $H(z)$ of Problems 6 and 12.

14. As in Chapter 1, §5.4, for moment systems we have, in general, the equations

$$H\Omega = 0, \qquad xD\Omega = 0$$

for the vacuum. Referring to §§2.3, 2.4, show that if Ω satisfies the above equations, then $\eta_n = R_t^n \Omega$ is a basis for an HW representation with raising operator R_t, lowering operator D, and satisfying $\eta_n = e^{tH} x^n \Omega$.

15. Referring to Proposition 3.1.3, check associativity of the radial translation $x \oplus y$ directly.

16. a. Let X be a random variable with density $p(x)$. Let a be a constant. Find $\langle f(X \oplus a) \rangle$ for the functions $f(x) = x$, $f(x) = x^2$.

   b. Discuss the possibility of defining variance in this context.

17. Calculate explicitly $\langle h(X \oplus Y) \rangle$ for $h(x) = x^2$, where X and Y have uniform distributions on $[0,1]$, i.e., X, e.g., has density $p(x) = 1$ for $x \in [0,1]$.

18. Referring to Proposition 3.1.1.1, show that if

$$\phi(z) = \sum_{n=0}^{\infty} \frac{z^n}{n! (c)_n} \mu_n$$

   Then μ_n are given by

$$\mu_n = (\Delta_z)^n \phi(z)\big|_{z=0}$$

   where $\Delta_z = z (d/dz)^2 + c \, d/dz$.

19. a. Prove Proposition 3.2.1.

   b. Show directly using the expressions given there that R_t, Δ_t, ρ_t indeed satisfy the sl(2) commutation rules.

20. Prove Proposition 3.2.2.

21. a. Calculate the Laguerre moment polynomials $l_n(x, c)$ for $n = 0, 1, \ldots, 6$.

   b. Calculate the corresponding moment polynomials according to Theorem 3.2.1.3 for the gamma density $(t > 0)$

$$p_t(x) = x^{t-1} e^{-x} / \Gamma(t), \qquad x \geq 0$$

22. Referring to Proposition 3.2.1.4, for your results of Problem 21a, verify that $y^n l_n(x/y, c)$ satisfy the evolution equation $\partial u / \partial y = \Delta u$, with $\Delta = x (\partial/\partial x)^2 + c (\partial/\partial x)$.

23. Do analogous problems to Problems 21 and 22 for the Bessel system, i.e., with the Hankel moment polynomials. Here the evolution equation becomes the generalized wave equation:

$$\Delta_y u = \Delta_x u$$

   where, e.g., $\Delta_y = y (\partial/\partial y)^2 + c (\partial/\partial y)$.

24. Give details of the proof of Proposition 3.3.1 according to the indications given there.

25. Check Proposition 3.3.4, #1, for the results of 21a, for $n = 1, 2, 3, 4$.

26. Verify the scaling property claimed in the remark following Proposition 3.3.5, that $h_n(x, t) = t^{n/2} h_n(x/\sqrt{t}, 1)$.

27. Verify Propositions 3.3.6, 3.3.8 for $n = 1, 2, 3$.

28. a. Write out the proof of Proposition 3.3.7.

    b. Check explicitly for $n = 2, 3$.

29. a. Verify Proposition 3.3.9 for $n = 0, 1, 2, 3$.

    b. Verify Proposition 3.3.10 for $n = 0, 1, 2, 3$.

30. Check that Definition 4.2 agrees with $\phi(D)x^n$ as claimed.

31. Work out the expansions corresponding to Theorem 4.3.2 for the following functions $V(z)$: i) $e^z - 1$ ii) $1 - e^{-z}$ iii) $z(1 - z)^{-1}$.

32. Fill in the details for Proposition 4.4.1.

33. As in Theorem 4.4.2.1, find $\exp(t\eta)$ for $\eta = xz(1 - z)$.

34. Discuss Theorem 4.4.3.2 for the gamma densities

$$p_x(y) = y^{x-1} e^{-y} / \Gamma(x), \qquad y \geq 0$$

35. Find the addition formulas corresponding to Proposition 4.4.4.1 for the following functions $V(z)$: i) $1 - e^{-z}$ ii) ze^{-z} iii) $z(1 - z)^{-1}$. In each case, for $n = 1, 2, 3$, write out the formulas explicitly.

36. Consider the uniform probability distribution on the numbers $0, 1, 2, \ldots, r - 1$. I.e., the probability of any of the numbers 0 through $r - 1$ is $1/r$. For $r = 6$ this corresponds to tossing a die. So we may call these measures generalized dice.

    a. Write the corresponding measure $p(dx)$ in the form $p(dx) = \sum p_n \delta_{a_n}(dx)$.

    b. Let $X_1, \ldots, X_N$ be independent with the same distribution, uniform on integers 0 to $r - 1$. Find the moment generating function for X_1 and for the sum $S_N = X_1 + \cdots + X_N$.

    c. Find the mean and variance of S_N.

37. a. Study the moment system corresponding to the distribution in the above Problem.

b. Let $H(z)$ (=log of the moment generating function of the distribution) be the generator for the moment system. Let $V(z) = H'(z) - \mu z$, where μ is the mean. Study the canonical system associated to V: $\xi_n(x)$. Use techniques related to Lagrange inversion (§4.4.1) to study the inverse function $U(v)$. (Try these for small values of r first, e.g., $r = 2, 3, 4$.)

5.2 QUANTUM HARMONIC OSCILLATOR (BIS)

Here we continue from Chapter 1, §5.3. Recall the raising operator $R = (x - D)/\sqrt{2}$ with the vacuum $\Omega = \exp(-x^2/2)$. Now we complete the calculation of the wave functions $\psi_n(x)$.

1. The operator $x - D$ can be expressed as

$$x - D = e^{-D^2/2}\, x\, e^{D^2/2}$$

2. Use the convolution integral

$$e^{D^2/2} f(x) = \int_{-\infty}^{\infty} f(x+y) e^{-y^2/2}\, dy/\sqrt{2\pi}$$

and the moment generating function for the Gaussian distribution p_t with $t = 1/2$, to find

$$e^{D^2/2}\, e^{-x^2/2} = e^{-x^2/4}/\sqrt{2}$$

3. Now we consider $(x - D)^n \Omega$. Via the sl(2) commutation rules find

$$(x - D)^n \Omega = e^{-D^2/2}\, e^{-x^2/4}\, x^n/\sqrt{2} = e^{-x^2/2}\, 2^\rho e^{-D^2}\, x^n/\sqrt{2}$$

with $\rho = xD + \frac{1}{2}$.

4. From the properties of Hermite polynomials deduce

$$(x - D)^n \Omega = 2^{n/2} H_n(x\sqrt{2}) e^{-x^2/2}$$

where $H_n(x) = H_n(x, 1)$.

5. Conclude that

$$\psi_n(x) = R^n \Omega = H_n(x\sqrt{2}) e^{-x^2/2}$$

Check with the results obtained previously for $\psi_1, \psi_2, \psi_3, \psi_4$.

5.3 SYMMETRIC FUNCTIONS AND MOMENT SYSTEMS

In this set of examples we will look at some basic features regarding the representations of the *symmetric group* S_N (the *permutation group* on N letters) and the theory of symmetric functions from the point of view of moment systems.

Given variables $\{x_1, \ldots, x_N\}$, the corresponding *homogeneous symmetric functions* and *elementary symmetric functions* may be defined via the generating functions

$$e^{\Lambda(z)} = \mathcal{E}(z) = \prod_1^N (1 - x_i z)^{-1} = \sum_{n \geq 0} z^n h_n(x_1, \ldots, x_N)$$

$$e^{-\Lambda(-z)} = \mathcal{E}^{-1}(-z) = \prod_1^N (1 + x_i z) = \sum_{n \geq 0} z^n a_n(x_1, \ldots, x_N)$$

where the function $\Lambda(z)$ is analogous to the generator $H(z)$ of the moment systems. We define the *power sums*

$$p_n = \sum_{i=1}^N x_i^n$$

1. Show that the exponent

$$\Lambda(z) = -\sum_i \log(1 - x_i z) = \sum_{n \geq 1} p_n \frac{z^n}{n}$$

5.3.1 Raising and lowering operators

2. Show that the function $\Lambda'(z) = \sum_n p_{n+1} z^n$ is the raising operator, i.e., multiplying by it is the same as sending $h_n \to (n+1)h_{n+1}$ (acting on the generating function $\mathcal{E}(z)$).

3. Show that the lowering operator is given by multiplication by z, sending $h_n \to h_{n-1}$.

4. Let $H_n = n! \, h_n$, then

$$\mathcal{E}(z) = \sum_{n=0}^{\infty} \frac{z^n}{n!} H_n$$

Indicating this correspondence by $\mathcal{E} \leftrightarrow \{H_n\}$, show that we have

$$\mathcal{E}' \leftrightarrow \{H_{n+1}\} \qquad z\mathcal{E} \leftrightarrow \{nH_{n-1}\}$$

à la ordinary differentiation $x^n \to nx^{n-1}$.

5.3.2 Characters of the symmetric group

The *character* χ_ρ^λ equals the trace of the representation λ for matrices corresponding to group elements in conjugacy class ρ. They are labelled by partitions of N, as follows. (See Ledermann[33], Macdonald[36])

ρ: denotes a conjugacy class in S_N with cycle structure ρ

λ: denotes a *tableau* or diagram corresponding to an irreducible representation of S_N of degree (i.e. dimension) d_λ

We will see that, actually, the characters mediate between the power sums and the h's and a's.

5.3.3 Fourier transform on the symmetric group

In general, given any class function $f(\rho)$ (i.e., a function on conjugacy classes, in the case of the symmetric group, a function of ρ), we define the 'Fourier transform'

$$\hat{f}(\lambda) = \frac{1}{N!} \sum_\rho |\rho| f(\rho) \chi_\rho^\lambda$$

5. Show that $|\rho|$, the number of elements of cycle type ρ, where $\rho = (1^{a_1} 2^{a_2} \cdots N^{a_N})$, is given by

$$|\rho| = \frac{N!}{1^{a_1} 2^{a_2} \cdots N^{a_N} a_1! a_2! \cdots a_N!}$$

The 'Fourier inversion formula' takes the form

$$f(\rho) = \sum_\lambda \hat{f}(\lambda) \chi_\rho^\lambda$$

6. Use the inversion formula to derive the *Plancherel identity*

$$\sum_\lambda \hat{f}(\lambda)^2 = \frac{1}{N!} \sum_\rho |\rho| f(\rho)^2$$

5.3.4 Schur functions

For each partition λ, we define the corresponding S-function (Schur function) by

$$\{\lambda\} = \det(h_{\lambda_i - i + j})$$

with the definitions $h_n = 0$ for $n < 0$ and $h_0 = 1$.

7. a. Show, for $N = 2$, $\lambda = [11]$, that

$$\{11\} = xy$$

b. Show that

$$\{2\} = h_2 = x^2 + y^2 + xy$$

Note that $\chi_\rho^{[N]}$ is the *unit character* the character of the trivial representation, mapping the whole group S_N to 1. The character, $\chi_\rho^{[1^N]}$, for the partition consisting of N 1's, is the *alternating character* the representation which sends a permutation to its *sign* +1 or −1 according to whether it is even or odd.

8. Check that for $N = 2$, $\chi_\rho^{[2]} = 1$ and

$$\chi_\rho^{[11]} = \begin{cases} -1, & \text{for } \rho = (2) \\ +1, & \text{for } \rho = (11) \end{cases}$$

5.3.5 Fundamental theorem for symmetric functions

Let, for cycle type $\rho = (1^{a_1} 2^{a_2} \cdots N^{a_N})$, the power sum functions p_ρ be defined by

$$p_\rho = p_1^{a_1} p_2^{a_2} \cdots p_N^{a_N}$$

The basic fact is:

The Fourier transform of the class function $f(\rho) = p_\rho$ is the S-function $\{\lambda\}$

The inversion formula says: $p_\rho = \sum_\lambda \{\lambda\} \chi_\rho^\lambda$. Via Newton's theorem that all symmetric functions may be expressed in terms of the power sums, we see that the S-functions $\{\lambda\}$ are a linear basis for symmetric functions homogeneous of degree N.

5.3.6 Conjugate relations

Let $\bar{h}_n = a_n$. The conjugate partition $\bar{\lambda}$ is defined as the partition corresponding to the transpose of the diagram of λ.

9. Show that for $[\lambda] = [3^2 2^3 1]$, with diagram

$$
\begin{array}{ccc}
\times & \times & \times \\
\times & \times & \times \\
\times & \times & \\
\times & \times & \\
\times & \times & \\
\times & &
\end{array}
$$

the transpose $[\bar{\lambda}]$ is $[652]$.

The S-functions satisfy the relation

$$
\{\bar{\lambda}\} = \det(\bar{h}_{\lambda_i - i + j}) = \det h_{\bar{\lambda}_i - i + j}
$$

$$
= \frac{1}{N!} \sum_\rho |\rho| \, p_\rho \chi_\rho^{\bar{\lambda}}
$$

and $\chi_\rho^{\bar{\lambda}} = \text{sgn}\,(\rho)\,\chi_\rho^{\lambda}$, where $\text{sgn}\,(\rho)$ is the alternating, or sign, character.

5.3.7 Relations with moment systems

Given the analytic function $\Lambda(z)$, we may define the functions h_n by the relation $\sum z^n h_n = e^{\Lambda(z)}$ and the conjugate functions $\bar{h}_n$ by $\sum \bar{h}_n = e^{-\Lambda(-z)}$. And the p_n are determined by $\sum z^n p_{n+1} = \Lambda'(z)$ or $\sum \frac{z^n}{n} p_n = \Lambda(z)$. Then the relationships given above among the functions h_n, a_n, p_n, and $\{\lambda\}$ will hold.

10. If we take $[\lambda] = [N]$, then

$$
h_N = \frac{1}{N!} \sum_\rho |\rho| \, p_\rho
$$

$$
a_N = \bar{h}_N = \frac{1}{N!} \sum_\rho |\rho| \, p_\rho \, \text{sgn}\,(\rho)
$$

5.3.8 Connection with moment systems

Now we can see the connection with HW representations and GMS. Set $x = p_1$ and scale $p_n \to t p_n$ so that $\Lambda(z) = xz + t \sum_{n=2}^{\infty} z^n (p_n / n)$.

11. Show that for

$$
\mathcal{E} = e^{\Lambda(z)} = e^{xz + t \sum_2^\infty z^n (p_n / n)}
$$

we have

$$\frac{\partial \mathcal{E}}{\partial t} = H(D)\,\mathcal{E}$$

with $D = d/dx$ and the generator $H(z) = \sum\limits_{n=2}^{\infty} z^n(p_n/n)$.

If we have a measure p, then, setting

$$H(z) = \log \int_{-\infty}^{\infty} e^{zx}\, p(dx)$$

with, say, $\int x\, p(dx) = 0$, $\int x^2 p(dx) = t$, we get

$$\mathcal{E} = \sum_{n=0}^{\infty} \frac{z^n}{n!}\, \eta_n(x,t)$$

and $\eta_n/n!$ correspond to the symmetric functions h_n.

12. Show that

$$\mathcal{E}^{-1}(-z) = \sum_{n=0}^{\infty} \frac{(-z)^n}{n!}\, \eta_n(-x,-t)$$

so that $\bar{\eta}_n(x,t) = (-1)^n \eta_n(-x,-t)$.

We need not specialize only p_1. Here we treat a single p_n as t.

13. a. Show that $n\frac{\partial}{\partial p_n}$ multiplies by z^n.

b. Deduce $n\frac{\partial}{\partial p_n} = (\frac{\partial}{\partial x})^n$.

Thus, we get a family of generalized heat equations:

$$\frac{\partial u}{\partial p_n} = \frac{1}{n}\frac{\partial^n u}{\partial x^n}$$

For the heat equation, §3.3, we have $\frac{\partial}{\partial t} = \frac{1}{2}(\frac{\partial}{\partial x})^2$, so $p_1 = x$, $p_2 = t$, with the rest of the p's zero.

We can define the *Airy* moment system, $\frac{\partial}{\partial t} = \frac{1}{3}(\frac{\partial}{\partial x})^3$, so $p_1 = x$, $p_3 = t$, with the other p's zero.

As usual, we have the raising operator $\Lambda'(z) = x + tH'(D)$ and $\eta_n = (x + tH')^n 1$.

The lowering operator is $z = D = \frac{d}{dx}$.

One can calculate the characters via this operator calculus, and, conversely, the characters yield the $\{\lambda\}$'s directly.

For S_3 the character table $\{\chi_\rho^\lambda\}$ is

$$
\begin{array}{c|ccc}
\lambda\backslash\rho & (3) & (21) & (111) \\
[3] & 1 & 1 & 1 \\
[21] & -1 & 0 & 2 \\
[111] & 1 & -1 & 1
\end{array}
$$

14. a. Show that corresponding to the usual heat equation:

$$p_1 = x, p_2 = t$$

$$3! \{ 3 \} = 3xt + x^3$$
$$3! \{ 21 \} = 2x^3$$
$$3! \{ 111 \} = -3xt + x^3$$

b. For the Airy system:

$$p_1 = x, p_3 = t$$

$$3! \{ 3 \} = 2t + x^3$$
$$3! \{ 21 \} = -2t + 2x^3$$
$$3! \{ 111 \} = 2t + x^3$$

5.4 HANKEL AND MELLIN TRANSFORMS

Here we study briefly Hankel and Mellin transforms. For more details on Hankel transforms see Lebedev[32]. For more on the Mellin transform, which plays an essential rôle in analytic number theory, e.g., see Rademacher[39].

5.4.1 Mellin transform

Define the *Mellin transform* of a function f (bounded at zero and decaying rapidly at infinity) by

$$M f(s) = \int_0^\infty x^{s-1} f(x)\, dx$$

for $\mathrm{Re}\, s > 0$, And, similarly, for a probability measure, we have the transform

$$\phi(s) = \int_0^\infty x^{s-1} p(dx)$$

which, for $p \in \mathcal{P}^*$, certainly exists for $\mathrm{Re}\, s \geq 1$.

The Mellin transform is related to the multiplicative group of positive real numbers, much as the Fourier transform is related to the additive group of real numbers.

1. Show that $Mf(s)$ is the Fourier-Laplace transform of $f(e^z)$.

2. Let $g(x) = f(\lambda x)$, $\lambda > 0$. Show that

$$Mg(s) = \lambda^{-s} Mf(s)$$

3. Let $\nu = xD$ denote the number operator. Using $\lim_{x \to \infty} f(x)x^s = 0$, find $M(\nu f)(s)$.

Now we look at the associated convolution structure.

4. Let $\phi_1(s)$, $\phi_2(s)$ be Mellin transforms of the probability measures p_1, p_2 respectively, corresponding to independent random variables X, Y.

   a. Show that $\phi_1(s) = \langle X^{s-1} \rangle$.

   b. Show that the product $\phi_1(s)\phi_2(s)$ is the Mellin transform of the distribution of the product XY.

5. Let $p_1(x)$, $p_2(y)$ be densities corresponding to the probability measures p_1, p_2 respectively (cf. above problem). Find the formula for the convolution $p_1 \times p_2$ corresponding to the relation

$$\int_0^\infty \int_0^\infty (xy)^{s-1} p_1(x) p_2(y)\, dx\, dy = \int_0^\infty u^{s-1}(p_1 \times p_2)(u)\, du$$

The Mellin transform is important in number theory because of the connection with Dirichlet series, which we indicate briefly.

6. Let $\mu(dx) = \sum a_n \delta_n(dx)$ be a, not necessarily finite, measure on the integers $\{n \geq 1\}$. Check the *Laplace transform* of μ

$$\int_0^\infty e^{-\lambda x} \mu(dx) = \sum_{n=1}^\infty a_n e^{-\lambda n}$$

7. Show that the Mellin transform of the Laplace transform of μ is the *Dirichlet series*

$$\Gamma(s) \sum_{n=1}^\infty \frac{a_n}{n^s}$$

8. Deduce that the Mellin transform of the Laplace transform of the 'uniform distribution' on $\{n \geq 1\}$ is $\Gamma(s)\zeta(s)$, where $\zeta(s)$ is the Riemann zeta function.

5.4.2 Hankel transform

Define the *Hankel transform* of a function f defined on the positive half-line by

$$H f(\lambda) = \int_0^\infty J_0(\lambda r) f(r) \, r \, dr$$

This comes from the Fourier transform of radial functions on $\mathbf{R}^2$. Let, the dot denoting the scalar product on $\mathbf{R}^2$,

$$\tilde{f}(y) = \int \int e^{i y \cdot x} f(x) \, dx$$

Then, changing to polar coordinates, we have, for a radial function f,

$$\tilde{f}(y) = \int_0^{2\pi} \int_0^\infty e^{i|y|r \cos \theta} f(r) \, r \, dr d\theta$$

1. Show that

$$\int_0^{2\pi} e^{it \cos \theta} \, d\theta = 2\pi \sum_{n=0}^\infty \frac{(-1)^n (t/2)^{2n}}{n! \, n!}$$

This last summation defines the *Bessel function $J_0(t)$* .

2. We thus have

$$\tilde{f}(|y|) = 2\pi \, H f(|y|) = 2\pi \int_0^\infty J_0(|y|r) \, f(r) \, r \, dr$$

3. From the formula

$$f(x) = \frac{1}{(2\pi)^2} \int \int e^{-i y \cdot x} \, \tilde{f}(y) \, dy$$

deduce the inversion formula for the Hankel transform

$$f(r) = \int_0^\infty J_0(r\lambda) \, H f(\lambda) \, \lambda \, d\lambda$$

4. For the Gaussian distribution with density

$$p_t(x, y) = e^{-(x^2 + y^2)/2t} / (2\pi t)$$

(the distribution of two independent Gaussian random variables with mean 0 and variance t) we have the transform

$$\int_0^\infty J_0(|y|r) \, e^{-r^2/2t} \, r \, dr = t e^{-|y|^2 t/2}$$

We can find the product formula for the function J_0.

5. In terms of the function $\mathcal{I}_c$, we have $J_0(t) = \mathcal{I}_1(-t^2/4)$.

6. Deduce an integral formula for the product $J_0(r) J_0(s)$ from the result for $\mathcal{I}_1$. (see Chapter 3, §4.3)

Chapter 5   BERNOULLI PROCESSES

This Chapter and the next present the main application of the theory developed in this text — the algebraic structure underlying the basic distributions and processes of probability theory. In this Chapter, first we present the Bernoulli systems based on the underlying Lie algebra structures. Then we discuss the associated stochastic processes: Bernoulli processes. In Chapter 6, various aspects of the analytic structure of Bernoulli systems are considered.

I.   Bernoulli systems: general structure

We will combine the time evolution from the theory of GMS with the Fock space to construct stochastic processes based on the HW, oscillator, and sl(2) algebras. One has the basis for the Fock space given by GMS polynomials with the time-zero representation given by canonical variables V, ξ. As these evolve in time, they yield systems of orthogonal polynomials for the corresponding probability measures. These are our *Bernoulli systems* .

Remark. This first section, §I, will present the general structure of Bernoulli systems. It is intended as a reference guide for the rest of the Chapter.

1.1   GMS AND CANONICAL VARIABLES

We will consider GMS, see Ch. 4, §2.4 for summary of properties, with the time-zero operators given by canonical variables V, ξ (Ch. 4, §IV). In the Fock space, we have an orthogonal basis. Recall that in Ch. 4, §3.3 the Hermite and Laguerre polynomials are related to moment systems by *time-reversal*. This is the basic technique. The following proposition explains that it is at least a step in the right direction.

1.1.1 Proposition.   Let $h_n(x, -t)$ be the GMS polynomials with generating function $\exp(ax - tH(a))$, corresponding to the time-reversed evolution. Suppose $\exp(tH(a)) = \int e^{ax} p_t(dx)$, for the convolution semigroup p_t. Then, with X_t denoting the corresponding random variables,

$$\langle e^{aX_t - tH(a)} \rangle = 1$$

and hence, $\langle h_n(X_t, -t) \rangle = 0$, for $n > 0$.

Proof: This follows directly by dividing the relation

$$e^{tH(a)} = \int_{-\infty}^{\infty} e^{ax} p_t(dx)$$

through by $\exp(tH(a))$. ∎

So, in general, the polynomials $h_n(x, -t)$ are orthogonal to $h_0 = 1$. We will see that this is the most one can expect in general. To get a full orthogonal basis requires the Fock space structure.

Let us see some features of GMS based on canonical variables V, ξ.

1.1.2 Definition. Given $H(z)$ the generator for a GMS and the canonical variable $V(z)$, with $U(v)$ the inverse of V, the *canonical generator* is $M(v) = H(U(v))$.

1.1.3 Proposition. With the time-zero basis $\{\xi_n(x)\}$, the generating function for the GMS polynomials is

$$e^{-tH} e^{v\xi} 1 = e^{xU(v) - tM(v)}$$

under the time-reversed flow.

Proof: In general, we have

$$e^{-tH} e^{ax} = e^{ax - tH(a)}$$

From Ch. 4, Theorem 4.2.1, we have the generating function for the time-zero basis

$$e^{v\xi} 1 = e^{xU(v)}$$

Substituting $a = U(v)$ yields the result. ∎

1.1.4 Definition. A *canonical GMS* is a GMS associated with canonical variables V, ξ, with the time-zero basis $\{\xi_n(x)\}_{n \geq 0}$, under the time-reversed evolution with generator $H(z)$.

We summarize some properties of a canonical GMS.

1. The canonical variables are V, ξ, with $V(z)$ a given function analytic at 0, $V(0) = 0$, $V'(0) \neq 0$, with corresponding inverse $U(v)$. With $W(z) = 1/V'(z)$, $\xi = xW(z)$.

2. The time-zero polynomials are $\xi_n(x) = \xi^n 1$, $n \geq 0$.

3. The initial raising and lowering operators are ξ and V. For $t > 0$, we drop the t subscript on R. We have the raising and lowering operators given by the Heisenberg flow:

$$R = e^{-tH} \xi e^{tH} = e^{-tH} xW(D)e^{tH} = (x - tH'(D))W(D)$$
$$V = e^{-tH} V e^{tH} = V(D)$$

4. The generating function for the system at time $t > 0$ is

$$e^{xU(v)-tM(v)} = \sum_{n=0}^{\infty} \frac{v^n}{n!} \xi_n(x, -t)$$

with the canonical generator $M(v) = H(U(v))$.

5. The polynomial $\xi_n(x, -t)$ is the solution to the (time-reversed) evolution equation:

$$\frac{\partial u}{\partial t} + H(D)u = 0, \qquad u(x, 0) = \xi_n(x)$$

and the generating function satisfies

$$\frac{\partial u}{\partial t} + H(D)u = 0, \qquad u(x, 0) = e^{v\xi} 1 = e^{xU(v)}$$

1.2 X OPERATOR AND FOCK SPACE

We give the 'general discussion', for sl(2), since we will see that the HW and oscillator algebra structures arise via limiting procedures (discussed in detail through the Chapter).

Remark. For the rest of this work, we will be considering vector spaces over **R**, real inner product spaces, and real Hilbert spaces.

Take the standard basis $\{\Delta, R, \rho\}$. Let

$$X = R + N + L$$

with $R = R$, $N = \alpha\rho$, $L = \beta\Delta$, acting on the basis $\psi_n = R^n \Omega$. On the vacuum state, Ω, $\rho\Omega = c\Omega$ and $L\Omega = 0$. The action on the basis is (cf. Ch. 1, Props. 3.3.1.1, 3.3.1.2 — note that this is a *different* scaling from Prop. 3.5.3; see also Ch. 3, eq. (2.4.1)):

$$R\psi_n = \psi_{n+1}, \qquad N\psi_n = \alpha(c + 2n)\psi_n, \qquad L\psi_n = \beta n(c + n - 1)\psi_{n-1} \qquad (1.2.1)$$

The letters R, N, L stand for raising, neutral, and lowering, respectively. With L and R adjoints, we have the squared norms

$$\gamma_n = \|\psi_n\|^2 = \beta^n n! (c)_n$$

according to Ch. 3, Prop. 2.2.1.1, cf. Ch. 3, Prop. 2.4.1. The idea is that since L, R are adjoints, and $N = (\alpha/\beta)[L, R]$ is self-adjoint, $X = R + N + L$ will be a self-adjoint operator. We want $X = X^*$ to give a *real* random variable. I.e., we will realize X as the operator of multiplication by x, the identity function on **R**, associated to a probability measure as in Ch.3, Theorem 1.4.2.

We define expectations in terms of the GNS form (Ch.3, §1.5) with respect to the vacuum state. Thus, for an operator Q on the Fock space,

$$\langle Q \rangle = \langle Q\Omega, \Omega \rangle$$

We normalize $\|\Omega\| = 1$. In the realization as functions, $\Omega = 1$, the constant function equal to 1, and the normalization corresponds to the fact that we are working with probability measures: $\langle 1 \rangle = 1$.

Now we can find the moment generating function of X.

1.2.1 Theorem. *The operator* $X = R + \alpha\rho + \beta\Delta = R + N + L$ *has moment generating function*

$$\langle e^{zX} \rangle = e^{cH(z)}$$

with $H(z) = \log(\delta \operatorname{sech} \delta z / (\delta - \alpha \tanh \delta z))$.

Proof: We use the splitting formula, Ch. 1, Prop. 4.3.1 to find:

$$\langle e^{zX} \Omega, \Omega \rangle = \langle e^{V(z)R} e^{\rho H(z)} e^{V(z)L} \Omega, \Omega \rangle$$

Since $L\Omega = 0$, and R is the adjoint of L, we have

$$\langle e^{V(z)R} e^{\rho H(z)} \Omega, \Omega \rangle = \langle e^{\rho H(z)} \Omega, e^{V(z)L} \Omega \rangle$$
$$= e^{cH(z)} \|\Omega\|^2 = e^{cH(z)}$$

■

This says that, p_c denoting the distribution of X,

$$\langle e^{zX} \rangle = \int_{-\infty}^{\infty} e^{zx} p_c(dx) = e^{cH(z)}$$

From our study of moment systems, we recognize that c is the *time-parameter for a flow corresponding to the convolution semigroup of measures* p_c. For this reason, from now on

we replace c by t

Thus, e.g., for the squared norms we write $\gamma_n = n! \beta^n(t)_n$.

Remark. Recall that c can be a negative integer (see the remark following Ch. 3, Prop. 2.4.1). This corresponds to the binomial process, the random walk of §II below. In such a case, we denote discrete time $-t$ by N. For the negative binomial process of §IV, we have discrete time N, in the infinite-dimensional case with $N = t > 0$. For all other processes we consider, t runs through the nonnegative reals, $[0, \infty)$.

Thus, as t varies, the operators X give a family of random variables X_t, forming a stochastic process with corresponding distributions p_t, a convolution semigroup of measures. Thus, X_t is a *time-homogeneous process with independent increments* (Ch. 4, §1.4.1). The discrete-time processes (binomial and negative binomial) are *random walks*. We will look at details in succeeding sections.

1.3 REALIZATION AS SPACES OF ORTHOGONAL POLYNOMIALS

Now, recall eq. (1.2.1), here replacing c by t:

$$R\psi_n = \psi_{n+1}, \qquad N\psi_n = \alpha(t+2n)\psi_n, \qquad L\psi_n = \beta n(t+n-1)\psi_{n-1} \qquad (1.3.1)$$

Thus the operator X_t is given via the action

$$X_t\psi_n = \psi_{n+1} + \alpha(t+2n)\psi_n + \beta n(t+n-1)\psi_{n-1} \qquad (1.3.2)$$

With $L\Omega = 0$, this gives $\psi_1 = (X_t - \alpha t)\Omega$. Thus, recursively, ψ_n are polynomials in the operator X_t, denoted by

$$\psi_n = J_n(X_t, t)\Omega$$

Taking $\Omega = 1$, writing $X_t = x$, the identity function on $\mathbf{R}$, we have the realization of the sl(2) Fock space as functions with basis

$$\psi_n = J_n(x, t)$$

Since the basis ψ_n is orthogonal, this is a family of orthogonal polynomials. These systems of polynomials are called *Bernoulli systems*, the terminology inspired by the Bernoulli distribution, Ch. 3, §1.3, that we will see is the foundation of these systems. We summarize the above discussion with

1.3.1 Proposition. *Bernoulli systems* $\{\, J_n(x, t)\,\}$ *satisfy the three-term recurrence*

$$xJ_n = J_{n+1} + \alpha(t+2n)J_n + \beta n(t+n-1)J_{n-1}$$

with $J_0 = 1$, $J_1 = x - \alpha t$. *The squared norms are*

$$\gamma_n = \|J_n\|^2 = n!\,\beta^n(t)_n$$

1.4 TIME EVOLUTION. BERNOULLI SYSTEMS AS CANONICAL GMS

We return to the splitting formula, as in the proof of Theorem 1.2.1, applied to Ω:

$$e^{zX_t}\,\Omega = e^{V(z)R}\,e^{tH(z)}\,\Omega$$

In the realization $X_t \to x$, $\Omega \to 1$, we have

$$e^{zx} = e^{tH(z)}\,e^{V(z)R}\,1$$

That is

$$e^{V(z)R}\,1 = e^{zx - tH(z)}$$

and we recognize the time-reversed flow of §1.1. Thus, with $z = U(v)$, inverse to $V(z)$, we have

$$e^{vR}\,1 = \sum_{n=0}^{\infty} \frac{v^n}{n!}\,R^n 1 = \sum_{n=0}^{\infty} \frac{v^n}{n!}\,J_n(x, t) \qquad (1.4.1)$$

$$= e^{xU(v) - tM(v)}$$

with canonical generator $M(v)$. That is, Bernoulli systems are canonical GMS. Thus, in terms of the canonical variables (z, x):

1.4.1 Proposition.   *The raising operator for Bernoulli systems, $RJ_n = J_{n+1}$, has the form*

$$R = xW(z) - tH'(z)W(z)$$

1.4.1 The canonical variable V. HW representation

Since these are GMS, we have the HW representation

$$RJ_n = J_{n+1}, \qquad VJ_n = nJ_{n-1}$$

Recall the proof of the splitting formula, Ch. 1, Prop. 4.3.1, and recall the remark there.

1.4.1.1 Definition.   The *characteristic polynomial* of the Bernoulli system with parameters α, β is

$$\pi(v) = 1 + 2\alpha v + \beta v^2$$

1.4.1.2 Proposition.   *The operator V is given as the solution to the Riccati equation*

$$V' = 1 + 2\alpha V + \beta V^2$$

with initial condition $V(0) = 0$. That is, V satisfies $V' = \pi(V)$.

We solve the equation for R given in Prop. 1.4.1, for x as the operator X_t:

$$X_t = tH' + R\pi(V) \tag{1.4.1.1}$$

Thus,

$$\frac{dX_t}{dt} = H'(z)$$

This is thus the *velocity* of the process in operator form. Now, from the proof of the splitting formula, we have the relation

$$H' = \alpha + \beta V \tag{1.4.1.2}$$

whence the terminology: V is the velocity operator.

Remark. In physics, Hamilton's equations give the relation

$$\frac{dx}{dt} = \frac{\partial H}{\partial p}$$

where $H = H(x, p)$ is the *Hamiltonian*, and p is the *momentum*. In our context, this gives the interpretation of the generator $H(z)$ as the Hamiltonian or *energy* and z as momentum.

To summarize, we give the form of the operators R, N, L in terms of the variables V, R, and we check the consistency of equations (1.3.2) and (1.4.1.1).

1.4.1.3 Proposition.   *In HW variables, we have*

$$R = R, \qquad N = \alpha(t + 2RV), \qquad L = \beta(tV + RV^2)$$

The operator X_t has the form

$$X_t = tH' + R\pi(V)$$

where $\pi(v)$ is the characteristic polynomial for the system.

Proof: This follows directly from eq. (1.3.1). For the operator X_t, via eq. (1.4.1.2):

$$\begin{aligned}
X_t &= tH' + R\pi(V) \\
&= t(\alpha + \beta V) + R(1 + 2\alpha V + \beta V^2) \\
&= R + \alpha(t + 2RV) + \beta(tV + RV^2) = R + N + L
\end{aligned}$$

as required.   ■

1.5   CHARACTERISTICS OF BERNOULLI SYSTEMS. GENERAL FORMS

Now for the description of these systems. We define the *characteristics*   of a given system:

1. The generator $H(z)$
2. The canonical variable (velocity operator) $V(z)$
3. The inverse function $U(v)$
4. The canonical generator $M(v)$
5. The canonical variable $\xi = xW(z)$, with $W(z) = 1/V'(z)$.

We are also interested in the number operator, RV, for these systems, since the polynomials $J_n(x, t)$ are eigenfunctions of the number operator, satisfying

$$RV J_n = n J_n$$

And we want to see the explicit form of the generating function and the polynomials $J_n(x, t)$.

1.5.1 General form of characteristics

Here we give the 'general formulas'. Then, in the rest of the chapter we present the interpretation of the associated stochastic processes and give the specific characteristics for each system.

The 'general Bernoulli system' comes from the sl(2) operator $X = R + N + L = R + \alpha\rho + \beta\Delta$. As above, we use the results from the splitting formula, here expressed in a form that fits well with the probabilistic interpretation.

1.5.1.1 Theorem.   *Given the parameters α, β, $\delta^2 = \alpha^2 - \beta$. The characteristics are given by:*

1. *The generator*
$$H(z) = \log \frac{2\delta}{(\delta - \alpha)e^{\delta z} + (\delta + \alpha)e^{-\delta z}}$$

2. *The canonical variable V*
$$V(z) = \frac{e^{\delta z} - e^{-\delta z}}{(\delta - \alpha)e^{\delta z} + (\delta + \alpha)e^{-\delta z}}$$

3. *The operators V and H are related by*
$$H'(z) = \alpha + \beta V(z)$$

4. *The inverse function $U(v)$*
$$U(v) = \frac{1}{2\delta} \log \frac{1 + (\alpha + \delta)v}{1 + (\alpha - \delta)v}$$

5. *The canonical generator*
$$M(v) = \tfrac{1}{2} \log \pi(v)$$

   where $\pi(v)$ is the characteristic polynomial $\pi(v) = 1 + 2\alpha v + \beta v^2$.

6. *The canonical variable ξ*
$$\xi = xW(z) = x \, (2\delta)^{-2}((\delta - \alpha)e^{\delta z} + (\delta + \alpha)e^{-\delta z})^2$$

Proof: The first two relations come directly from the splitting formula. We remarked relation #3 in eq. (1.4.1.2) above. The inverse function $U(v)$ is found by solving $V(U(v)) = v$, via #2. To see #5, we use Ch. 1, eq. (4.3.2), from the proof of the splitting formula, that $V' = \exp(2H)$. The Riccati equation for V says that $V' = \pi(V)$. Thus,
$$e^{2H(z)} = \pi(V(z))$$

Substituting $z = U(v)$, with $M(v) = H(U(v))$, the result follows. For #6, again use Ch.1, eq. (4.3.2), so that $W = 1/V' = \exp(-2H)$. Now the form of H given in #1 yields the result.   ∎

1.5.1.2 Corollary.   *The process X_t satisfies*
$$\langle X_t \rangle = \alpha t, \qquad \sigma^2(X_t) = \beta t$$

i.e., X_t has mean αt and variance βt.

Proof: We have $V(0) = 0$ and from the Riccati equation, $V'(0) = 1$. Now apply Ch. 4, Prop. 1.2.2. For X_1, we have by #3 of the Theorem, $\langle X_1 \rangle = \alpha$, $\sigma^2(X_1) = \beta$. For X_t, scaling H by t yields the result. ∎

Remark. We see that α is the *drift coefficient* giving the rate of deterministic motion, in this case, straight-line motion. And β is the *diffusion coefficient* controlling the fluctuations of the process about the mean motion.

We can write V and W nicely in terms of finite-difference operators.

1.5.1.3 Definition. The *finite-difference operator* given by $\partial_h = (e^{hD} - 1)/h$ acts on functions as

$$\partial_h f(x) = \frac{f(x+h) - f(x)}{h}$$

Remark. For $h > 0$, this is called a *forward difference operator*. Noting that $\partial_{-h} f(x) = (f(x) - f(x-h))/h$, this is a *backward difference operator*, for $h > 0$.

Rewriting the expressions for $V(z)$ and $W(z)$ given in Theorem 1.5.1.1, we have

1.5.1.4 Proposition. The operators V and W may be expressed in terms of finite-difference operators

$$V(z) = \frac{\partial_{2\delta}}{1 + (\delta - \alpha)\partial_{2\delta}}, \qquad W(z) = (1 + (\delta - \alpha)\partial_{2\delta})(1 - (\delta + \alpha)\partial_{-2\delta})$$

with $\partial_h = (e^{hz} - 1)/h$.

1.5.2 General forms for the orthogonal polynomial systems

The theory of GMS gives us the raising operator and the generating function. From the generating function we can find the form of the polynomials $J_n(x, t)$.

1.5.2.1 Theorem. For the general Bernoulli system:

1. The generating function $G_t(v, x) = \sum (v^n/n!) J_n(x, t)$ is given by

$$G_t(v, x) = (1 + (\alpha + \delta)v)^{(x-\delta t)/2\delta} (1 + (\alpha - \delta)v)^{-(x+\delta t)/2\delta}$$

2. The raising operator has the form

$$R = (x - \alpha t)W - \beta t V W$$

which can be expressed in terms of finite-difference operators as

$$R = [(x - \alpha t)(1 - (\delta + \alpha)\partial_{-2\delta}) - \beta t \partial_{-2\delta}](1 + (\delta - \alpha)\partial_{2\delta})$$

3. The number operator has the form

$$RV = (x - \alpha t)\partial_{-2\delta} + (\delta - \alpha)(x + \delta t)\partial_{2\delta}\partial_{-2\delta}$$

in terms of finite-difference operators.

4. The polynomials $J_n(x, t)$ have the form

$$J_n(x, t) = (\alpha + \delta)^n (-1)^n (t)_n \, {}_2F_1\left(\begin{matrix} -n, (x + \delta t)/2\delta \\ t \end{matrix} \, \middle| \, \frac{2\delta}{\alpha + \delta} \right)$$

Proof: Use the form

$$G_t(v, x) = e^{xU(v) - tM(v)}$$

from the theory of GMS, eq. (1.4.1), and apply Theorem 1.5.1.1. To substitute for $M(v)$, we observe that the characteristic polynomial factors as

$$\pi(v) = (1 + (\alpha + \delta)v)(1 + (\alpha - \delta)v)$$

Thus

$$e^{-tM(v)} = \pi(v)^{-t/2} = [(1 + (\alpha + \delta)v)(1 + (\alpha - \delta)v)]^{-t/2}$$

and, substituting in for $U(v)$, the result follows. For the raising operator, first we use Prop. 1.4.1 and #3 of Theorem 1.5.1.1. To get the expression in terms of finite-difference operators, use Prop. 1.5.1.4. For the number operator, RV, we use #2 and the expressions given in Prop. 1.5.1.4. It is convenient to use the relation

$$2\delta \, \partial_{2\delta}\partial_{-2\delta} = \partial_{2\delta} - \partial_{-2\delta}$$

Last, the explicit form of the polynomials $J_n(x, t)$ follows from the generating function, #1, and Ch. 2, Prop. 2.1. In the generating function for $_2F_1$ polynomials, make the substitutions: $x \to (\alpha + \delta)v$, $m \to (x - \delta t)/2\delta$, $a \to (\alpha - \delta)/(\alpha + \delta)$, and $n \to -(x + \delta t)/2\delta$. We observe that it is not necessary for m, n to be integers for the proof to hold, just that v be sufficiently close to the origin for the binomial expansions to be valid. ∎

1.5.3 The name 'Bernoulli'

A Bernoulli distribution, Ch. 3, §1.3, is a probability measure composed of two point masses. Let the points be $\mp\delta$ with respective probabilities $(1/2) \pm (\alpha/2\delta)$. Denoting the corresponding random variable by X, the expected value of $f \in C(\mathbf{R})$ is

$$\langle f(X) \rangle = (\tfrac{\delta + \alpha}{2\delta}) f(-\delta) + (\tfrac{\delta - \alpha}{2\delta}) f(\delta)$$

Now compute the moment generating function:

$$\langle e^{zX} \rangle = \tfrac{\delta + \alpha}{2\delta} e^{-\delta z} + \tfrac{\delta - \alpha}{2\delta} e^{\delta z}$$

Compare this with the formula for $H(z)$ given in Theorem 1.5.1.1. The Bernoulli distribution has moment generating function $\exp(-H(z))$. Recall, remarks following Ch. 3, Prop. 2.4.1 and Theorem 1.2.1 above, that for finite-dimensional sl(2) Fock spaces, t is a negative integer, $t = -N$. The process X_N is the sum of independent, identically distributed Bernoulli random variables, a random walk as discussed in Ch. 4, §1.4.1. Hence, applying Ch. 4, Prop. 1.3.3

$$\langle e^{zX_N} \rangle = e^{-NH(z)} = e^{tH(z)}$$

In order for these measures to be valid probability distributions, we must have $|\alpha/2\delta| < \frac{1}{2}$, or $|\alpha| < |\delta|$. With $\beta = \alpha^2 - \delta^2$, we see that this is just the condition that $\beta < 0$. The various processes discussed in this Chapter are distinguished by various choices of signs: $+$, $-$, 0 of these parameters. As each parameter α, β, or δ, goes to zero, the corresponding systems converge. We will see that these are given, and in fact give, interpretations of the classical elementary limit theorems of probability theory. With $\beta \to 0$, the Lie algebra changes from sl(2) to the oscillator ($\alpha \neq 0$) or HW algebra ($\alpha = \beta = 0$). This is called a *group contraction* and provides an algebraic interpretation of the limit theorems (Poisson limit theorem, central limit theorem).

II. Binomial process and Krawtchouk polynomials

The binomial process is the simplest random walk on the line. At each unit of time it jumps right or left. Each jump follows a Bernoulli distribution, the same at every step.

2.1 BERNOULLI RANDOM WALKS

We first want to find the moment generating function for the process X_N, the sum of N independent copies of a Bernoulli distribution. Each jump has the same distribution as X_1, which we assume takes values ± 1 with probabilities p, q respectively. Thus, the random walk moves one unit right or left at each unit of time.

2.1.1 Proposition.   *The moment generating function of X_N is*

$$\langle e^{zX_N} \rangle = (pe^z + qe^{-z})^N$$

where X_N is the Bernoulli random walk with jumps $+1$, -1 and corresponding probabilities p, q.

    Proof: The moment generating function of X_1 is $pe^z + qe^{-z}$ (cf., Ch. 3, §1.3). The result follows by Ch. 4, Prop. 1.3.3.   ∎

2.1.2 Proposition.   *X_N has a binomial distribution, with*

$$P(X_N = 2k - N) = \binom{n}{k} p^k q^{N-k}$$

so X_N takes values in $\{ -N, 2-N, \ldots, N-2, N \}$

Proof: Expanding by the binomial theorem:

$$\langle e^{zX_N} \rangle = (pe^z + qe^{-z})^N = \sum_{k=0}^{n} \binom{n}{k} e^{(2k-N)z} p^k q^{N-k}$$

as required. ■

Thus, this is the *binomial process* . Now we want to see the canonical GMS and associated structure.

2.2 CHARACTERISTICS FOR BINOMIAL PROCESS

From the moment generating function we see that, with N as time parameter, the generator of the binomial process is

$$H(z) = \log(pe^z + qe^{-z})$$

However, recalling that for finite-dimensional spaces, $t = -N < 0$, to agree with the general theory, we must consider $H(z) = -\log(pe^z + qe^{-z})$. Comparison with Theorem 1.5.1.1 gives us

2.2.1 Proposition. *The parameters for the binomial process are*

$$\alpha = 1 - 2p, \qquad \beta = -4pq, \qquad \delta = 1$$

From the general forms we have

2.2.2 Proposition. *The characteristics for the binomial process:*

$$H(z) = -\log(pe^z + qe^{-z}), \qquad M(v) = \tfrac{1}{2}\log\left[(1 - 2pv)(1 + 2qv)\right]$$
$$V(z) = \frac{(e^z - e^{-z})/2}{pe^z + qe^{-z}}, \qquad U(v) = \tfrac{1}{2}\log\frac{1 + 2qv}{1 - 2pv}$$

and $\xi = x(pe^z + qe^{-z})^2$.

The characteristic polynomial is $\pi(v) = 1 + 2(1 - 2p)v - 4pqv^2$.

2.3 ASSOCIATED GMS FOR THE BINOMIAL PROCESS

From the characteristics we can find the generating function and the orthogonal polynomials.

2.3.1 Proposition.   For the binomial process, with $t = -N$,

1. *The generating function*

$$G_t(v, x) = (1 + 2qv)^{(N+x)/2}(1 - 2pv)^{(N-x)/2}$$

2. *The orthogonal polynomials*

$$J_n(x, N) = 2^n K_n((N - x)/2, N) = N^{(n)}(2q)^n \, {}_2F_1\left(\begin{array}{c} -n, (x - N)/2 \\ -N \end{array} \middle| \frac{1}{q}\right)$$

   given in terms of Krawtchouk polynomials.

3. *The time-zero polynomials are given by*

$$\xi_n(x) = 2^n n! \sum_k \binom{-x/2}{k}\binom{x/2}{n-k}(-1)^k p^k q^{n-k}$$

Proof: These follow directly from the general forms. We take the form of the Krawtchouk polynomials given in the reference list, Intro., cf., Ch. 3, Def. 3.1.1.3. For #3, use the generating function with $N = 0$ directly.   ■

The raising operator and number operator are most readily given by the general forms, Theorem 1.5.2.1.

2.4 LIE STRUCTURE AND RECURRENCE RELATIONS, BINOMIAL PROCESS

From the characteristic polynomial we have

2.4.1 Proposition.   For the binomial process, the operator $X = R + N + L$ corresponds to the recurrence relation

$$xJ_n = J_{n+1} + (1 - 2p)(2n - N)J_n + 4pqn(N - n + 1)J_{n-1}$$

with $J_0 = 1$, $J_1 = x + (1 - 2p)N$. The squared norms are

$$\gamma_n = n! \, (4pq)^n N^{(n)}, \qquad 0 \le n \le N$$

We find for the polynomials K_n:

2.4.2 Proposition.   *The Krawtchouk polynomials satisfy the recurrence*

$$(qN - x)K_n = K_{n+1} + (q - p)nK_n + pqn(N - n + 1)K_{n-1}$$

with $K_0 = 1$, $K_1 = qN - x$.

Proof: Use the relation $K_n(x, N) = 2^{-n}J_n(N - 2x, N)$ in the above Proposition.   ∎

Since here we are following the general forms, the Lie algebra is sl(2). It is important to note that these representations, being finite-dimensional, correspond to the compact Lie group SU(2). Thus, these are the irreducible representations of su(2), basic to the theory of spin in quantum theory. (See Biedenharn-Louck[6])

2.5 SYMMETRIC BINOMIAL PROCESS

An important special case arises when $p = q = \frac{1}{2}$. Then the random walk is *symmetric*, as likely to move east as to go west. The formulas specialize nicely so that they afford a good illustration of the general theory as well.

Properties of symmetric binomial systems:

1. We have the parameters $\alpha = 0$, $\beta = -1$, $\delta = 1$.
2. The generator is $H(z) = -\log\cosh z$, with $t = -N$. And $V(z) = \tanh z$, $W(z) = \cosh^2 z$, $U(v) = \frac{1}{2}\log((1 + v)/(1 - v))$.
3. The characteristic polynomial is $\pi(v) = 1 - v^2$.
4. The generating function for the polynomials is

$$G_t(v, x) = (1 + v)^{(N+x)/2}(1 - v)^{(N-x)/2}$$

   with the orthogonal polynomials

$$J_n(x, N) = N^{(n)} {}_2F_1\left(\begin{matrix} -n, (x - N)/2 \\ -N \end{matrix} \middle| 2\right)$$

5. The raising operator $R = (x\cosh z - N\sinh z)\cosh z$.
6. The number operator $RV = (x\cosh z - N\sinh z)\sinh z$.

Remark. The operators $\cosh z$ and $\sinh z$ act via unit shifts:

$$(\cosh D)f(x) = \frac{f(x + 1) + f(x - 1)}{2}, \qquad (\sinh D)f(x) = \frac{f(x + 1) - f(x - 1)}{2}$$

III.   Negative binomial process and Meixner polynomials

Here we encounter the notion of *space–time dual* . Namely, if we have a process X_t, we can consider a dual process T_x, defined as the amount of time required for the process X_t to first reach the position x. Necessarily $T_x > 0$, for x other than the starting point, which we conventionally take to be zero.

3.1   NEGATIVE BINOMIAL AS A TIME PROCESS

The *negative binomial distribution* is the distribution of the space-time dual of a binomial process with Bernoulli increments taking values 0 and 1, with probabilities q and p respectively. I.e., the binomial process either sticks or it jumps to the right one step. Meanwhile, the negative binomial process counts time waiting for the binomial process to reach a given position. So here $\{X_N = k\}$ means that it takes k steps for the binomial process to reach N.

3.1.1 Proposition.   *For the negative binomial distribution:*

1. *The moment generating function is*

$$\langle e^{zX_N} \rangle = \left(\frac{pe^z}{1 - qe^z} \right)^N$$

2. *The probabilities are given by*

$$P(X_N = k) = \binom{k-1}{N-1} p^N q^{k-N}, \qquad \text{for } k \geq N$$

Proof: We have only to find the distribution of X_1, as X_N is as sum of N independent, identically distributed copies of X_1. By construction, we see that X_1 has a *geometric distribution*

$$P(X_1 = k) = pq^{k-1}, \qquad k \geq 1$$

since necessarily the first jump occurred at the kth time unit. The moment generating function is given by, via geometric series,

$$\langle e^{zX_1} \rangle = \frac{pe^z}{1 - qe^z}$$

This gives #1. Expanding this result yields the probabilities as stated in #2.   ∎

3.2 CHARACTERISTICS FOR NEGATIVE BINOMIAL

From the moment generating function, we see that

$$\tfrac{1}{N} \log\langle e^{zX_N} \rangle = \log \frac{pe^z}{1 - qe^z}$$

does not quite match the general form in Theorem 1.5.1.1. What we will do is to shift back by a half-step. That is, replace X_1 by $X_1 - \tfrac{1}{2}$. We then have:

$$\langle e^{zX_1} \rangle = \frac{pe^{z/2}}{1 - qe^z}$$

so we are counting $X_1 = \tfrac{1}{2}, \tfrac{1}{2} + 1, \tfrac{1}{2} + 2, \dots$. And we have

$$H(z) = \log \frac{1}{p^{-1}e^{-z/2} - qp^{-1}e^{z/2}}$$

3.2.1 Proposition.   *The parameters for the negative binomial process are*

$$\alpha = p^{-1} - \tfrac{1}{2}, \qquad \beta = qp^{-2}, \qquad \delta = \tfrac{1}{2}$$

Proof: From Theorem 1.5.1.1, we see that $\delta = 1/2$. Then $\delta + \alpha = p^{-1}$, consistent with $\delta - \alpha = -qp^{-1}$, and $\beta = \alpha^2 - \delta^2$ follows.   ∎

Now the general forms give

3.2.2 Proposition.   *The characteristics for the negative binomial process*

$$H(z) = \log \frac{pe^{z/2}}{1 - qe^z}, \qquad M(v) = \tfrac{1}{2} \log \left[(1 + qp^{-1}v)(1 + p^{-1}v) \right]$$

$$V(z) = \frac{p(e^z - 1)}{1 - qe^z}, \qquad U(v) = \log \frac{1 + p^{-1}v}{1 + qp^{-1}v}$$

with $\xi = xp^{-2}e^{-z}(1 - qe^z)^2$.

The characteristic polynomial is $\pi(v) = 1 + (2p^{-1} - 1)v + qp^{-2}v^2$.

3.3 ASSOCIATED GMS FOR THE NEGATIVE BINOMIAL PROCESS

For the GMS we have

3.3.1 Proposition.   *For the negative binomial process, for the orthogonal polynomials we have:*

1. *Generating function*

$$G_t(v,x) = (1 + p^{-1}v)^{x-N/2}(1 + qp^{-1}v)^{-x-N/2}$$

2. *The orthogonal polynomials are **Meixner polynomials***

$$J_n(x,N) = M_n(x,N) = (-p)^{-n}(N)_n\,{}_2F_1\left(\begin{array}{c} -n, x + N/2 \\ N \end{array}\middle| p\right)$$

3. *The time-zero polynomials are*

$$\xi_n(x) = n!\, p^{-n} \sum_k \binom{x}{k}\binom{-x}{n-k} q^{n-k}$$

3.4 LIE STRUCTURE AND RECURRENCE RELATION, NEGATIVE BINOMIAL

Following the general theory, we have

3.4.1 Proposition.   *For the negative binomial process, the operator $X = R + N + L$ corresponds to the recurrence relation*

$$xM_n = M_{n+1} + (p^{-1} - \tfrac{1}{2})(N + 2n)M_n + qp^{-2}n(N + n - 1)M_{n-1}$$

with $M_0 = 1$, $M_1 = x - (p^{-1} - \tfrac{1}{2})N$. The squared norms are given by

$$\gamma_n = (qp^{-2})^n n!\,(N)_n$$

Here the Lie algebra sl(2) corresponds to the noncompact Lie group SU(1,1). Thus, these are discrete series representations of the Lie algebra su(1,1).

IV. Continuous binomial process and Meixner-Pollaczek polynomials

When $\delta^2 < 0$, i.e., δ is purely imaginary, leads to an interesting distribution. It is a continuous analogue of the binomial/negative binomial distributions. Since α and β are real parameters, $\delta^2 < 0$ implies $\beta = \alpha^2 - \delta^2 > 0$. So we choose a convenient parametrization:

$$\alpha = \tan \vartheta, \qquad \beta = \sec^2\vartheta, \qquad \delta = i$$

Here we are not beginning with a description of the process as it is not easily described. From the general theory of processes with independent increments, one knows that it moves in jumps of arbitrary size. It is an interesting process to study. Here we discuss some details of interest from the present point of view. (See Feller[20].)

4.1 CONTINUOUS BINOMIAL DISTRIBUTION

First we have a useful calculation:

4.1.1 Proposition.   *Let X be a random variable with distribution given by the beta density*

$$p_t(x) = \frac{2^{t-1}}{2\pi} B\left(\frac{t+ix}{2}, \frac{t-ix}{2}\right), \qquad t > 0$$

then the moment generating function of X is

$$\langle e^{zX} \rangle = (\sec z)^t$$

Proof: Start with the integral representation of the beta function, Intro., Prop. 3.2, #2:

$$B(b-a, a) = \int_0^\infty \frac{u^{a-1}}{(1+u)^b} \, du$$

The density function is thus:

$$p_t(x) = \pi^{-1} 2^{t-2} \int_0^\infty \frac{u^{\frac{t+ix}{2}-1}}{(1+u)^t} \, du$$

Substituting $u = e^{2y}$ gives

$$\frac{2^t}{2\pi} \int_{-\infty}^\infty \frac{e^{y(t+ix)}}{(1+e^{2y})^t} \, dy = \frac{2^t}{2\pi} \int_{-\infty}^\infty e^{iyx} \left(\frac{e^y}{1+e^{2y}}\right)^t dy$$

Since $p_t(x)$ and $\operatorname{sech} y$ are even functions, we have

$$p_t(x) = \frac{1}{2\pi} \int_{-\infty}^\infty e^{-iyx} (\operatorname{sech} y)^t \, dy$$

We will find the characteristic function

$$\phi(is) = \int_{-\infty}^\infty e^{isx} p_t(x) \, dx$$

By Fourier inversion this gives $\phi(is) = (\operatorname{sech} s)^t$ and hence the moment generating function as stated above. Here we indicate how to derive the inversion.

Consider, for $v > 0$,

$$\int_{-\infty}^\infty e^{isx - v|x|} p_t(x) \, dx = \frac{1}{2\pi} \int_{-\infty}^\infty \int_{-\infty}^\infty e^{i(s-y)x} e^{-v|x|} (\operatorname{sech} y)^t \, dx \, dy$$

Integrating out x, first for $x > 0$ and then for $x < 0$, gives

$$\frac{1}{2\pi} \int_{-\infty}^{\infty} \left[\frac{1}{v - i(s - y)} + \frac{1}{v + i(s - y)} \right] (\operatorname{sech} y)^t \, dy = \frac{v}{\pi} \int_{-\infty}^{\infty} \frac{(\operatorname{sech} y)^t}{v^2 + (s - y)^2} \, dy$$

Now check that

$$\frac{v}{\pi} \int_{-\infty}^{\infty} \frac{dy}{v^2 + y^2} = 1$$

and that, as $v \to 0^+$, it gives an approximate delta function, i.e.,

$$\frac{v}{\pi} \int_{|y| > \varepsilon} \frac{dy}{v^2 + y^2} \to 0$$

as $v \to 0^+$, for any $\varepsilon > 0$. Thus, taking $v \to 0$, using Lebesgue's dominated convergence theorem for the limit inside the integral, we get $(\operatorname{sech} s)^t$ and hence the result. ∎

We introduce ϑ as a shift: $z \to z + \vartheta$, $|\vartheta| < \pi/2$.

4.1.2 Corollary. *Let X have density*

$$p_t(x) = \frac{(2 \cos \vartheta)^t}{4\pi} e^{\vartheta x} B\left(\frac{t + ix}{2}, \frac{t - ix}{2} \right)$$

Then the moment generating function of X is $(\cos \vartheta / \cos(z + \vartheta))^t$.

Proof: We just have to note the normalization that at $z = 0$, the moment generating function gives the value 1. ∎

4.2 CHARACTERISTICS OF CONTINUOUS BINOMIAL PROCESS

Checking that indeed

$$H(z) = \log \frac{\cos \vartheta}{\cos(z + \vartheta)}$$

corresponds to the parameters $\alpha = \tan \vartheta$, $\beta = \sec^2 \vartheta$, $\delta = i$, we find

4.2.1 Proposition. *Characteristics for the continuous binomial process:*

$$H(z) = \log \frac{\cos \vartheta}{\cos(z + \vartheta)}, \qquad M(v) = \tfrac{1}{2} \log(1 + 2v \tan \vartheta + v^2 \sec^2 \vartheta)$$

$$V(z) = \frac{\tan z}{1 - \tan \vartheta \tan z}, \qquad U(v) = \arctan\left(\frac{v}{1 + v \tan \vartheta} \right)$$

with $\xi = x \cos^2(z + \vartheta) \sec^2 \vartheta$.

The characteristic polynomial is $\pi(v) = 1 + 2v \tan \vartheta + v^2 \sec^2 \vartheta$.

Remark. Note that we can write $M(v)$ as well in the form: $\log |1 + v(\tan \vartheta - i)|$.

4.3 ASSOCIATED GMS FOR THE CONTINUOUS BINOMIAL PROCESS

Here we have the *Meixner-Pollaczek polynomials* as the orthogonal basis. We denote them by $B_n(x,t)$, the B standing for binomial (and reminiscent of Bernoulli).

4.3.1 Proposition. *The orthogonal polynomials for the continuous binomial process have:*

1. *Generating function*

$$G_t(x,v) = \exp\left[x \arctan\left(\frac{v}{1+v\tan\vartheta}\right)\right] \times (1 + 2v\tan\vartheta + v^2\sec^2\vartheta)^{-t/2}$$

2. *The orthogonal polynomials have the form*

$$J_n(x,t) = B_n(x,t) = (-i\sec\vartheta)^n e^{-in\vartheta}(t)_n \, {}_2F_1\left(\begin{array}{c}-n, (t-ix)/2 \\ t\end{array}\middle| 1 + e^{2i\vartheta}\right)$$

3. *The time-zero polynomials are*

$$\xi_n(x) = n!\,(i\sec\vartheta)^n e^{-in\vartheta} \sum_k \binom{ix/2}{k}\binom{-ix/2}{n-k}(-1)^k e^{2ik\vartheta}$$

4.4 LIE STRUCTURE AND RECURRENCE RELATIONS, CONTINUOUS BINOMIAL

From the general theory we find

4.4.1 Proposition. *For the continuous binomial process, the operator* $X = R + N + L$ *corresponds to the recurrence relation*

$$xB_n = B_{n+1} + (\tan\vartheta)(t+2n)B_n + (\sec^2\vartheta)n(t+n-1)B_{n-1}$$

with $B_0 - 1$, $B_1 = x - t\tan\vartheta$. *The squared norms are*

$$\gamma_n = \sec^{2n}\vartheta\, n!\,(t)_n$$

Here we have infinite-dimensional representations with X having continuous spectrum. These correspond to the noncompact group $SL(2,\mathbf{R})$.

V.   Poisson process and Poisson-Charlier polynomials

This process and the two following arise via *limit theorems*. The exponential process, §VI, arises as well as a discrete-time process — as the space-time dual of the Poisson process. We will now see in what sense the Bernoulli distribution is fundamental. Namely, the Poisson and Brownian motion processes are continuous-time processes arising as limits of the binomial process by taking time increments $\Delta t \to 0$. Similarly, the exponential process, as the space-time dual of the Poisson process, arises as a corresponding limit of the negative binomial process, this being the space-time dual of the binomial process. These processes play important rôles in applications dealing with random phenomena, such as signals, that run in continuous time.

5.1 POISSON LIMIT THEOREM

We start with a binomial process, with Bernoulli-distributed increments, taking values $0, 1$, with probabilities q, p respectively. This is the 'stick-or-jump' process of §3.1. The idea is to take time steps Δt progressively shorter so that in the limit $\Delta t \to 0$, we have a continuous-time process that jumps at random.

First fix a time unit. Call each Bernoulli increment a 'trial' — do you stick or jump? Then consider the following schemes:

1. The binomial process makes one trial per unit time, with the probability of a jump on a given trial equal to p.

2. Now, take $\tau = (\Delta t)^{-1}$ trials per unit time, with probability of jump per trial equal to $\lambda \Delta t$, where $\lambda > 0$ denotes the average rate of jumps per unit time.

Let $X^{(\tau)}$ denote the number of jumps made during one time unit, with $\tau = (\Delta t)^{-1}$, i.e., there are τ trials per unit time. What is the moment generating function of $X^{(\tau)}$? We know that $X^{(\tau)}$ is binomial with $p = \lambda \Delta t$, for $N = \tau$ steps:

$$\langle e^{zX^{(\tau)}} \rangle = (qe^0 + pe^z)^\tau = \left(1 + \frac{\lambda}{\tau}(e^z - 1)\right)^\tau$$

As $\Delta t \to 0$, $\tau \to \infty$, and we have

$$\langle e^{zX^{(\infty)}} \rangle = e^{\lambda(e^z - 1)} = e^{-\lambda} \sum_{n=0}^{\infty} \frac{\lambda^n}{n!} e^{nz}$$

This is a Poisson distribution (Ch. 3, §1.3) with parameter λ. If we scale time by a factor $t > 0$, we have

$$\langle e^{zX_t} \rangle = e^{\lambda t(e^z - 1)}$$

with now X_t a *Poisson process* running in continuous time. Here X_t is the total number of jumps up to time t, with λ the average number of jumps per unit time.

A simple modification yields the moment generating function when the size of the jumps is α, a fixed number. We just scale $X \to \alpha X$.

5.1.1 Proposition. *The moment generating function for a Poisson process with rate λ and jump size α is*

$$\langle e^{zX_t} \rangle = e^{\lambda t(e^{\alpha z} - 1)}$$

the corresponding probability distributions are given by:

$$p_t(dx) = e^{-\lambda t} \sum_{n=0}^{\infty} \frac{(\lambda t)^n}{n!} \delta_{\alpha n}(dx)$$

Note that we can absorb λ into t as choice of scale. Thus, we put $\lambda = 1$ for the remainder of the discussion.

5.2 POISSON LIMIT THEOREM: LIE STRUCTURE AND FOCK SPACE

Now we want to see how the Lie algebra and Bernoulli systems structures correspond. First, consider a binomial process with jump size δ. I.e., recalling §1.5.3, the binomial process jumps $+\delta$ or $-\delta$ per unit time. To get a 'stick-or-jump' process we replace each increment, X, say, by $X + \delta$. This gives a stick-or-jump process with jump size 2δ. For N steps we have the process $Y_N = X_N + N\delta$. For the Bernoulli systems in general, we replace N by $-t$, giving the process Y_t:

$$Y_t = X_t - \delta t$$

Next, we look at the parameters α, β, δ. For the Poisson limit theorem, the jump size δ is fixed. From Prop. 2.2.1, we have, setting $p = \lambda/\tau$: $\alpha = 1 - 2(\lambda/\tau) \to 1$ and $\beta = -4\lambda\tau^{-1}(1 - \lambda/\tau) \to 0$, as $\tau \to \infty$. Thus, we look for the behavior of the Bernoulli systems as $\beta \to 0$. Since the number of steps per unit time scales as τ, going to infinity, we scale $t \to t/\beta$. Note that this is the scaling for L in Ch. 1, Prop. 3.5.2.

5.2.1 Theorem.   For the general Bernoulli process, consider the operator $X_t - \delta t$, with t scaled to t/β, in the limit $\beta \to 0$, i.e., in the *Poisson limit* . Then

$$X_{t/\beta} - \delta t/\beta \to R + 2\alpha RV + tV + t/2\alpha = X_{osc} + t/2\alpha$$

where X_{osc} denotes an operator from the oscillator algebra.

Proof: In HW variables, we have, Prop. 1.4.1.3,

$$X_t = R + \alpha(t + 2RV) + \beta(tV + RV^2)$$
$$X_{t/\beta} - \delta t/\beta = R + (\alpha - \delta)(t/\beta) + 2\alpha RV + tV + \beta RV^2$$

In the limit $\beta = (\alpha^2 - \delta^2) \to 0$, we take the choice of signs so that $\alpha - \delta \to 0$ and $\alpha + \delta \to 2\alpha$. And we have $(\alpha - \delta)/\beta \to (2\alpha)^{-1}$ as $\beta \to 0$. Thus, the result follows. ∎

So the Lie algebra associated to the Poisson process is the *oscillator algebra*. It is important to see how the operators H, V and the generating function G behave in this limit. For $Y_t = X_t - \delta t$, note that

$$\langle e^{zY_t} \rangle = \langle e^{zX_t} \rangle e^{-t\delta z}$$

and that the scaling $t \to t/\beta$ effectively scales H to $\beta^{-1}H$. Thus,

5.2.2 Proposition.   In the Poisson limit, $t \to t/\beta$, $\beta \to 0$, we have

1. For the generator
$$\beta^{-1}(H(z) - \delta z) \to (e^{2\delta z} - 1)/4\delta^2$$

2. The canonical variable
$$V(z) \to (e^{2\alpha z} - 1)/2\alpha$$
satisfying $V' = 1 + 2\alpha V$.

3. The generating function
$$G_t(x,v)\Big|_{x \to x + \delta t, t \to t/\beta} \to (1 + 2\alpha v)^{x/2\alpha} e^{-vt/2\alpha}$$

which, for $\alpha = 1/2$, is the generating function for the *Poisson-Charlier polynomials*.

Proof: For #1, we have

$$\beta^{-1}(H(z) - \delta z) = \log\left(\frac{\delta - \alpha}{2\delta} e^{2\delta z} + \frac{\delta + \alpha}{2\delta}\right)^{-1/\beta}$$

$$= \log\left(\frac{\delta - \alpha}{2\delta}(e^{2\delta z} - 1) + 1\right)^{-1/\beta}$$

$$= \log\left(1 - \frac{\beta}{2\delta(\alpha + \delta)}(e^{2\delta z} - 1)\right)^{-1/\beta}$$

$$\to (e^{2\delta z} - 1)/4\delta^2 \qquad \text{as } \beta \to 0$$

The proof of #2 is straightforward. Since $X_t = Y_t + \delta t$, we replace x by $x + \delta t$ in the generating function. Via Theorem 1.5.2.1,

$$G_t(x,v)\Big|_{x \to x + \delta t, t \to t/\beta} = (1 + (\alpha + \delta)v)^{x/2\delta}(1 + \beta(\alpha + \delta)^{-1}v)^{-x/2\delta - t/\beta}$$

$$\to (1 + 2\alpha v)^{x/2\alpha} e^{-vt/2\alpha}$$

as $\beta \to 0$.   ∎

Now let us look at Ch. 1, Prop. 4.2.1. We have R, V instead of x, D, acting on the HW Fock space, with normalized vacuum state Ω. Thus,

5.2.3 Proposition.   Consider the operator $X_t = R + N + L = R + \alpha RV + tV$ on the HW Fock space, with vacuum Ω, $L\Omega = N\Omega = 0$. Then, as a random variable, X_t has moment generating function

$$\langle e^{zX_t} \rangle = \exp\left[t(e^{\alpha z} - 1 - \alpha z)/\alpha^2\right]$$

i.e., X_t is a (centered and scaled) Poisson process.

Proof: Here, R and L are adjoints, with $L = tV$, so, by Ch. 1, Prop. 4.2.1:

$$\langle e^{zX_t} \rangle = \langle e^{zX_t} \Omega, \Omega \rangle$$
$$= e^{tH(z)} \langle e^{V(z)R} e^{zN} e^{V(z)L} \Omega, \Omega \rangle$$
$$= e^{tH(z)}$$

since $N\Omega = 0$ (and $\|\Omega\|^2 = 1$). ∎

Comparing with Theorem 5.2.1, we see that, with 2α for α in the above proposition,

$$\langle e^{z(X_t + t/2\alpha)} \rangle = \exp[t(e^{2\alpha z} - 1)/4\alpha^2]$$

where we recall that in the limit, $\alpha - \delta \to 0$, so here we have α replacing δ.

5.3 CHARACTERISTICS FOR POISSON PROCESS

For the 'standard' version of the Poisson process, we replace 2α by α. And we take X_t to be the centered and scaled process corresponding directly to the oscillator algebra. Now,

5.3.1 Proposition. *In the context of Bernoulli systems, the Poisson process corresponds to parameters $\alpha \neq 0$, $\beta = 0$.*

Observe that the equation $H' = \alpha + \beta V$ no longer holds, since it would imply that the process be deterministic. Here we have $H' = V$. And

5.3.2 Proposition. *The characteristics for the Poisson process*

$$H(z) = (e^{\alpha z} - 1 - \alpha z)/\alpha^2, \qquad M(v) = \alpha^{-1}v - \alpha^{-2}\log(1 + \alpha v)$$
$$V(z) = (c^{\alpha z} \quad 1)/\alpha, \qquad U(v) = \alpha^{-1}\log(1 + \alpha v)$$

with $\xi = xe^{-\alpha z}$.

The characteristic polynomial is $\pi(v) = 1 + \alpha v$.

Now we can discuss the GMS structure.

5.4 ASSOCIATED GMS FOR POISSON PROCESS. RECURRENCE

Here we have the HW-type Fock space with

$$R\psi_n = \psi_{n+1}, \qquad N\psi_n = \alpha n\psi_n, \qquad L\psi_n = tn\psi_{n-1} \qquad (5.4.1)$$

Since we are taking Ω to be a unit vector, we have, Ch. 3, Prop. 2.3.1, the squared norms

$$\gamma_n = n!\, t^n$$

In the realization with X_t given as the variable x, we have, as in §1.4, the generating function for the orthogonal polynomials $G_t(v, x) = \exp(xU(v) - tM(v))$ and, since here $H' = V$, the raising operator $R = (x - tV)W$.

5.4.1 Proposition.   For the Poisson process, in terms of finite-difference operators we have:

1. The operators V and W

$$V = \partial_\alpha, \qquad W = 1 - \alpha\partial_{-\alpha}$$

2. The raising operator is

$$R = x - (\alpha x + t)\partial_{-\alpha}$$

3. The number operator

$$RV = x\partial_{-\alpha} - t\partial_\alpha\partial_{-\alpha}$$

And for the orthogonal polynomials we have

5.4.2 Proposition.   For the Poisson process, we have

1. Generating function

$$G_t(x, v) = (1 + \alpha v)^{(x/\alpha + t/\alpha^2)}\, e^{-vt/\alpha}$$

2. The orthogonal polynomials are given in terms of Poisson-Charlier polynomials

$$J_n(x, t) = \alpha^n P_n(x/\alpha + t/\alpha^2, t/\alpha^2)$$

with

$$P_n(x, t) = (-t)^n\, {}_2F_0\left(\begin{array}{c} -n, -x \\ \underline{} \end{array}\middle|\, -t^{-1}\right)$$

3. The time-zero polynomials are *factorial polynomials*

$$\xi_n(x) = x(x - \alpha)\cdots(x - \alpha(n - 1))$$

Remark. As indicated above, Prop. 5.2.2, the Poisson-Charlier polynomials have the generating function

$$(1 + v)^x e^{-vt} = \sum_{n=0}^{\infty} \frac{v^n}{n!}\, P_n(x, t)$$

We have, via eq. (5.4.1),

5.4.3 Proposition.   The Poisson basis $\{J_n(x, t)\}$ satisfies the recurrence

$$xJ_n = J_{n+1} + \alpha n J_n + t n J_{n-1}$$

with $J_0 = 1$, $J_1 = x$.

With the relation $P_n(x, t) = J_n(x - t, t)$ for $\alpha = 1$,

5.4.4 Corollary.   The Poisson-Charlier polynomials satisfy the recurrence

$$xP_n = P_{n+1} + (t + n)P_n + t n P_{n-1}$$

with $P_0 = 1$, $P_1 = x - t$.

VI.   Exponential process and Laguerre polynomials

The exponential is a continuous-time process that arises naturally as a discrete-parameter process: the space-time dual of the Poisson process. We define

6.1 Definition.   The *gamma distributions* are given by density functions of the form

$$p_t(x) = x^{t-1}e^{-x} / \Gamma(t)$$

for $x \in [0, \infty)$, with $t > 0$.

Remark. In the above Definition, we are not including the parameter λ explicitly. It is common to define the gamma densities in general as

$$p_t(x) = \lambda^t x^{t-1} e^{-\lambda x} / \Gamma(t)$$

for $x \in [0, \infty)$, with $\lambda, t > 0$. For $t = 1$, we have the *exponential distribution* with parameter λ.

6.1   TIME PROCESS FOR POISSON. LIMIT THEOREM

As the space-time dual of the Poisson process, it is extremely simple to find the distribution of the exponential process. Namely, let Y_t be a Poisson process, with jumps of size 1. Let X_N be the space-time dual. Note that this is a discrete-parameter process. The event $\{X_N = s\}$ means that the Nth jump of Y_t occurs at time s.

6.1.1 Proposition.   *For the exponential process, X_N, as the space-time dual of the Poisson process:*

1. *X_1 has an exponential distribution with density*

$$p_1(x) = e^{-x}$$

   for $x \geq 0$.

2. *The moment generating function of X_N is $(1-z)^{-N}$.*

3. *The generator is $H(z) = -\log(1-z)$.*

Proof: Consider if the Nth jump of Y_t has not yet occurred:

$$P(X_N > s) = P(Y_s < N)$$

In particular, with $N = 1$:

$$P(X_1 > s) = P(Y_s < 1) = P(Y_s = 0) = e^{-s}$$

Thus, for $s > 0$,

$$\int_s^\infty p_1(dx) = e^{-s}$$

and differentiating yields #1. The calculation of the moment generating function of X_1 is immediate, and the moment generating function of X_N is the Nth power. The statement of #3 follows directly from #2. ∎

It is readily checked from the definition of the gamma function that if X_t has a gamma distribution, with parameter t as defined above, then the moment generating function of X_t is $(1-z)^{-t}$. Thus, the discrete-parameter exponential process extends to continuous-time, which we call the *exponential process* even though one could properly call it the gamma process.

Since the discrete-parameter exponential process is dual to the Poisson process, we expect that the negative binomial, as the space-time dual to the binomial process, converges accordingly in the Poisson limit. Since this is a transition from discrete to continuous time from the point of view of the binomial-to-Poisson limit, we expect the 'jump-size' of the time process to go to zero. Thus, we expect a scaling of z in the moment generating function.

6.1.2 Proposition.   *In the limit $\tau \to \infty$ with $p = \lambda \tau^{-1}$, and the scaling $z \to z\tau^{-1}$, we have*

$$\left(\frac{pe^z}{1 - qe^z} \right)^N \to \left(\frac{1}{1 - \lambda^{-1}z} \right)^N$$

That is, the distribution of X_N of the negative binomial process converges to the corresponding gamma distribution of the exponential process.

Proof: In $pe^z/(1 - qe^z)$, the moment generating function for the geometric distribution (negative binomial at time 1), substitute $p \to \lambda\tau^{-1}$ and $z \to z\tau^{-1}$. Then we have

$$\frac{pe^z}{1 - qe^z} = \frac{\lambda\tau^{-1}e^{z/\tau}}{1 - (1 - \lambda\tau^{-1})e^{z/\tau}}$$

$$= \frac{\lambda}{\lambda + \tau(e^{-z/\tau} - 1)} \to \frac{\lambda}{\lambda - z}$$

as $\tau \to \infty$. Multiplying numerator and denominator by λ^{-1}, we see that this gives the result for X_1. The result for X_N follows by taking N^{th} powers. ∎

This version, with λ retained explicitly, is consistent with the interpretation of λ^{-1}. Since λ is the average number of jumps per unit time of the Poisson process, i.e., the *frequency* of jumps, the reciprocal λ^{-1} is the *period*, i.e., the average waiting time between jumps. And it follows readily from the integral for the gamma function that, indeed, if Y has an exponential distribution with density $\lambda e^{-\lambda x}$ on $[0, \infty)$, maintaining λ explicitly, then $\langle Y \rangle = \lambda^{-1}$.

6.2 LIE STRUCTURE FOR EXPONENTIAL PROCESS

Here the scaling $z \to z/\tau$ with τ going to infinity indicates that it is a matter of the parameter δ going to zero. From Prop. 3.2.1, we see as well that, as $p \to 0$,

$$\frac{\alpha^2}{\beta} = \frac{(p^{-1} - 1/2)^2}{qp^{-2}} \to 1$$

consistent with their difference being bounded.

6.2.1 Proposition. *For the general Bernoulli process, for the operator X_t, as $\delta \to 0$:*

$$X_t \to R + \alpha(t + 2RV) + \alpha^2(tV + RV^2)$$

a special form of the sl(2) X operator.

Proof: This is clear from the relation $\delta^2 = \alpha^2 - \beta$. ∎

What is interesting here is the form of the operators H, V and the generating function G_t.

6.2.2 Theorem. *In the limit $\delta \to 0$, the general Bernoulli system yields for the exponential process:*

1. *The generator*

$$H(z) \to -\log(1 - \alpha z)$$

2. *The canonical variable*

$$V(z) \to \frac{z}{1 - \alpha z}$$

3. *The generating function*

$$G_t(x, v) \to (1 + \alpha v)^{-t} \exp\left(\frac{v}{1 + \alpha v} x\right)$$

which, for $\alpha = 1$, is the generating function for the Laguerre polynomials.

Proof: Using the form of $H(z)$ as given in the splitting formula, Ch. 1, Prop. 4.3.1, we have

$$\exp(H(z)) = \frac{\delta \operatorname{sech} \delta z}{\delta - \alpha \tanh \delta z} = \frac{1}{\cosh \delta z - \alpha \delta^{-1} \sinh \delta z}$$

$$\to (1 - \alpha z)^{-1}$$

For #2, use the form of $V(z)$ from the splitting formula in a similar way. For #3, rewriting the generating function, Theorem 1.5.2.1, #1, we have

$$\left(\frac{1 + \delta v(1 + \alpha v)^{-1}}{1 - \delta v(1 + \alpha v)^{-1}}\right)^{x/2\delta} (1 + \alpha v)^{-t} \to (1 + \alpha v)^{-t} \exp(xv(1 + \alpha v)^{-1})$$

as $\delta \to 0$. ■

Compare this generating function with that in Ch. 4, Prop. 3.3.4, noting the remark after Ch. 4, Def. 3.2.1.2.

6.3 CHARACTERISTICS FOR EXPONENTIAL PROCESS

First, we give the parameters.

6.3.1 Proposition.   *The exponential process is a Bernoulli process with $\delta = 0$, i.e., the relation $\alpha^2 = \beta$ holds.*

Now we have, from Theorem 6.2.2 and the above discussion,

6.3.2 Proposition.   *For the exponential process, the characteristics are:*

$$H(z) = -\log(1 - \alpha z), \qquad M(v) = \log(1 + \alpha v)$$
$$V(z) = \frac{z}{1 - \alpha z}, \qquad U(v) = \frac{v}{1 + \alpha v}$$

with $W(z) = (1 - \alpha z)^2$.

The characteristic polynomial is $\pi(v) = (1 + \alpha v)^2$.

6.4 GMS STRUCTURE FOR EXPONENTIAL PROCESS. RECURRENCE FORMULAS

Since this is a special form of the sl(2) algebra, we have an sl(2)-type Fock space. So let us proceed right away to give the raising operator and number operator.

6.4.1 Proposition.   *For the exponential process, in terms of differential operators*

1. *The raising operator is*

$$R = x(1 - \alpha z)^2 - \alpha t(1 - \alpha z)$$

2. *The number operator is*

$$RV = xz(1 - \alpha z) - \alpha t z$$

For the orthogonal polynomials, the generating function G_t is given above in Theorem 6.2.2.

6.4.2 Proposition.   *For the exponential process:*

1. *The orthogonal polynomials are given in terms of Laguerre polynomials:*

$$J_n(x,t) = \alpha^n L_n(x/\alpha, t)$$

with

$$L_n(x,t) = (-1)^n(t)_n \, {}_1F_1 \left(\begin{matrix} -n \\ t \end{matrix} \middle| x \right)$$

2. *The time-zero polynomials (with $\alpha = 1$) have the form, for $n > 0$,*

$$\xi_n(x) = \sum_{k=1}^{n} \binom{n}{k} \frac{\Gamma(n)}{\Gamma(k)} (-1)^{n-k} x^k$$

Proof: The result #1 follows from Ch. 4, Prop. 3.3.4, and the remark following Def. 3.2.1.2 of Chapter 4 (as remarked above regarding the generating function). Statement #2 follows by writing out the ${}_1F_1$ summation and setting $t = 0$. Observe that, for $n > 0$, the first term, with $k = 0$, drops out.   ∎

We have from the form of the operator X_t, Prop. 6.2.1,

6.4.3 Proposition.   *For the exponential process, the basis $\{\,J_n(x,t)\,\}$ satisfies the recurrence*

$$xJ_n = J_{n+1} + \alpha(t + 2n)J_n + \alpha^2 n(t + n - 1)J_{n-1}$$

with $J_0 = 1$, $J_1 = x - \alpha t$.

And, specifically,

6.4.4 Corollary.   *With $\alpha = 1$, the above relation gives the recurrence for the Laguerre polynomials $L_n(x,t)$.*

VII.   Brownian motion and Hermite polynomials

The stochastic process *Brownian motion*   plays a fundamental rôle in the mathematical theory of diffusion and stochastic differential equations. It appears extensively in many fields of pure and applied mathematics, physics, and engineering, especially, but not exclusively, in statistical physics and communication theory. An interesting application of the theory of diffusion processes to computer science will appear in Volume 2. The formal derivative of Brownian motion is *white noise* corresponding to a completely random hiss, with equal contributions from all parts of the frequency spectrum. (Some references: Kac[28], Breiman[10], Hida[26])

7.1 BROWNIAN MOTION AND THE CENTRAL LIMIT THEOREM

Here we give an approach that should give the reader some idea of what the process looks like. Start with the symmetric binomial process, a Bernoulli random walk, jumping either right or left one step with probability 1/2. Now we want to construct a continuous-time process. The main feature is the *fractal scaling*

$$(\Delta x)^2 \sim \Delta t$$

That is, as $\Delta t \to 0$, we maintain $(\Delta x)^2$ proportional to Δt. For standard Brownian motion, we take $(\Delta x)^2 = \Delta t$. The result of this is to create the effect of *diffusion* rather than a usual mechanical motion. For example, one expects the trajectories to have no defined velocity, which is indicated by the fact that, as distinguished from smooth curves:

$$\frac{\Delta x}{\Delta t} = \frac{1}{\sqrt{\Delta t}} \to \infty, \qquad \text{as } \Delta t \to 0$$

The 'jagged' nature of the random walk trajectories is thus maintained in the limit. The scheme is the following:

1. The binomial process $X^{(1)}$ takes one step of size 1 (right or left) in one time unit.

2. The binomial process $X^{(100)}$ takes 100 steps of size 1/10 in one time unit.

3. The process $X^{(\tau)}$ takes τ steps of size $\tau^{-1/2}$ in one time unit.

   The limiting process as $\tau \to \infty$ is Brownian motion.

The *central limit theorem* in general involves convergence of distributions to the normal or *Gaussian distribution* . The convergence of the scaled binomial distributions to the Gaussian distribution is known as the *De-Moivre–Laplace limit theorem*.

Note that the scaling $\Delta x = \tau^{-1/2}$ effectively scales the variable $z \to z/\sqrt{\tau}$:

$$\text{If } \phi(z) = \langle e^{zX} \rangle, \quad \text{then } \langle e^{zX/\sqrt{\tau}} \rangle = \phi(z/\sqrt{\tau})$$

Recalling §2.5,

7.1.1 Proposition.   With the scaling $\Delta t = 1/N$, $\Delta x = 1/\sqrt{N}$, the symmetric binomial random variables $X_1^{(N)}$ converge to X_1, with moment generating function

$$\langle e^{zX_1} \rangle = e^{z^2/2}$$

Thus, X_1 has a standard (i.e., centered and scaled) Gaussian distribution.

Proof: The symmetric binomial variable $X_1^{(N)}$ has moment generating function

$$\langle e^{z X_1^{(N)}} \rangle = 2^{-N}(e^{z/\sqrt{N}} + e^{-z/\sqrt{N}})^N$$

$$= \left(1 + \frac{z^2}{2N} + \mathcal{O}(N^{-3/2})\right)^N$$

$$\to e^{z^2/2}$$

as $N \to \infty$. ∎

Scaling by t, for the standard Brownian motion process we have

7.1.2 Proposition.   *For standard Brownian motion, the distribution of X_t is given by the density*

$$p_t(x) = \frac{e^{-x^2/2t}}{\sqrt{2\pi t}}$$

Proof: The scaling $x \to x\sqrt{t}$ shows that we can put $t = 1$. We have the moment generating function:

$$\int_{-\infty}^{\infty} e^{zx} e^{-x^2/2} \, dx/\sqrt{2\pi} = e^{z^2/2} \int_{-\infty}^{\infty} e^{-(z-x)^2/2} \, dx/\sqrt{2\pi}$$

For $z \in \mathbf{R}$, the integral is just $\int p_1(x)\, dx = 1$. For complex z, the result follows by analytic continuation (or via the integral by suitably applying Cauchy's theorem). ∎

7.2   LIE STRUCTURE FOR BROWNIAN MOTION

As we have seen in the limit theorems discussed so far, the various time-/space-scalings correspond to the vanishing of certain of the parameters α, β, or δ. Here we have both space and time scaling. As we have seen, recall §§1.5.3, 5.2, e.g., the *space scale* is given by δ and the *time scale* by β. In the symmetric binomial case, $\alpha = 0$. Thus, $\delta^2 = -\beta \to 0$.

7.2.1 Definition.   The *diffusion limit* is given by the scaling $t \to t/\beta$, and the limits α, $\beta \to 0$.

Of course, $\delta \to 0$ as well.

7.2.2 Theorem.   *For the general Bernoulli process, in the diffusion limit we have for the operator X_t:*

$$X_{t/\beta} \to R + tV = X_{HW}$$

where X_{HW} is the X-operator for the Heisenberg-Weyl algebra.

Proof: This follows directly from the general form

$$X_t = R + \alpha(t + 2RV) + \beta(tV + RV^2)$$

■

So we have come to the HW algebra. From Ch. 1, Prop. 4.1.1, we have

7.2.3 Proposition.   *On the HW Fock space, the operator* $X_t = R + tV$, *considered as a random variable, has moment generating function*

$$\langle e^{zX_t} \rangle = e^{z^2 t/2}$$

i.e., $\{X_t\}_{t \geq 0}$ *is a Brownian motion process.*

Let us look at the convergence of H, V, and the generating function G_t. Here we find the Hermite polynomials, see Ch 4., Prop. 3.3.1 and Def. 3.3.2, giving the orthogonal basis for the Fock space.

7.2.4 Theorem.   *In the diffusion limit, we have:*

1. *Convergence of the moment generating functions*

$$\exp(tH(z)) \to e^{z^2 t/2}$$

2. *Convergence of the canonical variable*

$$V(z) \to z$$

3. *The generating function*

$$G_t(v, x) \to e^{xv - v^2 t/2}$$

the generating function for the Hermite polynomials.

Proof: Let us take the limit from the symmetric binomial system, i.e., let $\alpha = 0$, $\beta = -\delta^2$. This gives

$$\exp(tH(z)) \Big|_{\substack{t \to t/\beta \\ \alpha \to 0}} = \left(\frac{1}{\cosh \delta z} \right)^{t/\beta}$$

$$= (\cosh \delta z)^{t/\delta^2} \to e^{z^2 t/2}$$

as in Prop. 7.1.1. Thus #1. For #2, we have

$$V(z) \Big|_{\alpha \to 0} = \delta^{-1} \tanh \delta z \to z, \qquad \text{as } \delta \to 0$$

And for the generating function:

$$G_t(x,v)\Big|_{\alpha \to 0} = \left(\frac{1+\delta v}{1-\delta v}\right)^{x/2\delta}(1-\delta^2 v^2)^{-t/2\beta}$$

$$= \left(1 + \frac{2\delta v}{1-\delta v}\right)^{x/2\delta}(1+\beta v^2)^{-t/2\beta}$$

$$\to e^{vx - v^2 t/2}$$

as $\delta \to 0$.   ∎

Remark. Starting with $\beta \to 0$, §§5.3, 5.4, one can similarly check that for the Poisson→Gaussian limit, $\alpha \to 0$, there is appropriate convergence of H, V, and G_t. Observe that, in Prop. 5.4.2, #1 and #2, the required relation $(\Delta x)^2 = \Delta t$ is evident in the scaling factors α^2 and α.

7.3 CHARACTERISTICS FOR BROWNIAN MOTION

From the point of view of the general Bernoulli systems, this is, of course, the simplest of the Bernoulli processes.

7.3.1 Proposition.   *The Brownian motion process corresponds to parameters* $\alpha = \beta = 0$.

As for the Poisson process, we have $H' = V$. And

7.3.2 Proposition.   *The characteristics for Brownian motion*

$$H(z) = z^2/2, \qquad M(v) = v^2/2$$
$$V(z) = z, \qquad U(v) = v$$

with $\xi = x$.

The characteristic polynomial is $\pi(v) = 1$.

Note that for Brownian motion, the canonical variables are the same as the usual (z, x). Here the interpretation of H as energy, z as momentum, and V as velocity is apparent. We recognize immediately $M(v)$ as the kinetic energy of a free particle of unit mass.

7.4 GMS STRUCTURE AND RECURRENCE FORMULA FOR BROWNIAN MOTION

Here we have the raising and number operators as differential operators.

7.4.1 Proposition.   *For Brownian motion*

1. *The raising operator is*
$$R = x - tz$$

2. *The number operator is*
$$RV = xz - tz^2$$

And for the orthogonal polynomials, Hermite polynomials, the generating function is given above, Theorem 7.2.4. As we saw in Ch. 4, §3.3, the Hermite polynomials are the time-reversed Hermite moment polynomials.

7.4.2 Proposition.   *For Brownian motion, we have*

1. *The orthogonal polynomials are given by Hermite polynomials:*

$$J_n(x,t) = H_n(x,t)$$

which may be expressed in terms of hypergeometric functions:

$$H_{2n}(x,t) = (-2t)^n \left(\tfrac{1}{2}\right)_n {}_1F_1 \left(\begin{matrix} -n \\ 1/2 \end{matrix} \,\middle|\, x^2/2t \right)$$

$$x^{-1} H_{2n+1}(x,t) = (-2t)^n \left(\tfrac{3}{2}\right)_n {}_1F_1 \left(\begin{matrix} -n \\ 3/2 \end{matrix} \,\middle|\, x^2/2t \right)$$

2. *The time-zero polynomials are the ordinary powers x^n.*

Proof: From the remark before Ch. 4, Prop. 3.3.6, we have

$$H_n(x,t) = i^n t^{n/2} h_n(x/i\sqrt{t})$$

with $h_n(x)$ the Hermite moment polynomials at time 1. Thus,

$$H_{2n}(x,t) = (-t)^n h_{2n}(x/i\sqrt{t})$$
$$\tfrac{i\sqrt{t}}{x} H_{2n+1}(x,t) = i^{2n+1} t^{n+1/2} \left(\tfrac{i\sqrt{t}}{x} h_{2n+1}(x/i\sqrt{t}) \right)$$

Now the result, #1, follows from Ch. 4, Prop. 3.3.9. And #2 is clear from the generating function.   ∎

The recurrence relation, via $X = R + tV$,

7.4.3 Proposition.   *The Hermite polynomials satisfy the recurrence relation*

$$xH_n = H_{n+1} + tnH_{n-1}$$

with $H_0 = 1$, $H_1 = x$.

VIII.   Canonical moments

We can think of the coefficients of the expansion of $\exp(tM(v))$, the exponential of the canonical generator, as moments for a *canonical measure* $p_t(d\xi)$. With the factorization

$$G(v) = e^{xU(v)} \pi(v)^{-t/2}$$

with the replacement $t \to -t$, these correspond to the $x = 0$ polynomials. Thus, we have the *canonical moments* with generating function

$$\pi(v)^{t/2} = \sum_{n=0}^{\infty} \frac{v^n}{n!}\, \tilde{\mu}_n(t)$$

Recall the generating function for Gegenbauer polynomials,

$$(1 - 2\alpha v + v^2)^{-\nu} = \sum_{m=0}^{\infty} v^m C_m^{\nu}(\alpha)$$

Comparison gives us:

8.1 Proposition.   *The canonical moments are:*

$$\tilde{\mu}_n(t) = (-1)^n n!\, \beta^{n/2} C_n^{-t/2}(\alpha/\sqrt{\beta})$$

The general form of the polynomials $J_n(x, -t)$ is

$$J_n(x, -t) = (\alpha + \delta)^n t^{(n)}\, {}_2F_1 \left(\begin{matrix} -n, (x - \delta t)/2\delta \\ -t \end{matrix} \ \middle| \ \frac{2\delta}{\alpha + \delta} \right) \qquad (8.1)$$

and the canonical moments come by evaluating at $x = 0$. Replacing $\alpha = \sqrt{\beta} \cosh \theta$, $\delta = \sqrt{\beta} \sinh \theta$ in eq. (8.1) yields:

$$\tilde{\mu}_n(t) = \beta^{n/2} e^{n\theta}\, t^{(n)}\, {}_2F_1 \left(\begin{matrix} -n, -t/2 \\ -t \end{matrix} \ \middle| \ 1 - e^{-2\theta} \right)$$

Remark. Comparing with the expression in terms of Gegenbauer polynomials gives, setting $\beta = 1$, $t = -2\nu$, and $\theta \to i\theta$:

$$C_n^{\nu}(\cos \theta) = e^{in\theta} \frac{(2\nu)_n}{n!}\, {}_2F_1 \left(\begin{matrix} -n, \nu \\ 2\nu \end{matrix} \ \middle| \ 1 - e^{-2i\theta} \right)$$

Remark. The standard reference on orthogonal polynomials is Szëgo[45]. The material for Chapter 5 is based on Feinsilver[16]. For connections with Lie groups in general, see Feinsilver&Schott[17], [18].

IX.   Exercises and examples

9.1 EXERCISES

1. As in Theorem 1.2.1, find the moment generating function for the operators $X = x + \alpha d$, $X = x + \alpha x D + \beta D$ coming from the HW and oscillator algebras respectively.

2. Use Proposition 1.3.1 to calculate J_2, J_3, J_4.

3. Prove Proposition 1.4.1 by differentiating equation (1.4.1) with respect to v.

4. Fill in the details of Theorem 1.5.1.1.

5. Referring to Definition 1.5.1.3, show that $\partial_h = (e^{hz} - 1)/h$ satisfies:

   a. $h\,\partial_h\partial_{-h} = \partial_h - \partial_{-h}$

   b. $\sinh hz \cosh hz = (h/2)(\partial_{2h} + \partial_{-2h})$

   c. $\sinh^2 hz = h^2\partial_{2h}\partial_{-2h}$

6. Verify Proposition 1.5.1.4.

7. Fill in the details of Theorem 1.5.2.1.

8. Verify the statements in §2.5.

9. Check the steps of the proof of Proposition 4.1.1.

10. Verify the claim regarding the form of $H(z)$ made at the beginning of §4.2.

11. In §§II–VII, verify the Propositions regarding the characteristics, GMS, etc., for each of the Bernoulli processes.

12. For each process, calculate the corresponding polynomials for $n = 2$ and 3 using the appropriate recurrence formula.

13. Show the recurrence formula for Gegenbauer polynomials $\{\, C_n^\nu(x)\,\}$.
    Let $\phi_n(x) = n!\,C_n^\nu(x)$. Then

$$2x(\nu + n)\phi_n = \phi_{n+1} + n(2\nu + n - 1)\phi_{n-1}$$

with $\phi_0 = 1$, $\phi_1 = 2\nu x$.

9.2 CANONICAL MEASURES

As mentioned in §VIII, we can determine a *canonical measure* $p_t(d\xi)$ by the relation

$$e^{tM(v)} = \int_{-\infty}^{\infty} e^{v\xi} p_t(d\xi)$$

where here ξ denotes simply the variable of integration. Let $v = iy$, for $y \in \mathbf{R}$. Then we have the Fourier transform pair

$$\int_{-\infty}^{\infty} e^{iy\xi} p_t(d\xi) = e^{tM(iy)}$$

and

$$p_t(\xi) = \int_{-\infty}^{\infty} e^{-iy\xi} e^{tM(iy)} \, dy/2\pi \qquad (9.2.1)$$

where $p_t(\xi)$ is the (possibly formal) density for the measure $p_t(d\xi) = p_t(\xi)d\xi$. Note that in canonical variables we have the generating function

$$G_t(v,x) = e^{-tM(V)} e^{v\xi} 1$$

We proceed to study the various Bernoulli systems.

1. From Proposition 7.3.2, conclude that there is no difference between the canonical measure and the usual Gaussian measure.

2. For the exponential system, referring to Proposition 6.3.2, with $\alpha = 1$, we see that $\exp(-tM(v))$ is the Fourier-Laplace transform of the distribution of $-X_t$, where X_t has a gamma distribution, i.e., this is the negative of the exponential process X_t. So the canonical process is a space-time reflection of the original process. Discuss the case $\alpha \neq 1$.

3. From Proposition 5.3.2, we have for the Poisson process with $\alpha = 1$

$$e^{tM(v)} = e^{vt} (1+v)^{-t}$$

which is the Fourier-Laplace transform for the distribution of $t - X_t$, where X_t is an exponential process. Discuss the case $\alpha \neq 1$.

For the Bernoulli systems corresponding to the various binomial processes, we require some integrals leading to Bessel functions. First we have

4. Check the relation, with $\pi(v) = 1 + 2\alpha v + \beta v^2$,

$$\pi(v) = \beta \left[\left(v + \frac{\alpha}{\beta} \right)^2 - \frac{\delta^2}{\beta^2} \right]$$

5. Verify the integral formula

$$\int_{-1}^{1} e^{su} \left(1 - u^2\right)^{t/2} du = B(1/2, 1 + t/2) \, \mathcal{I}_{(t+3)/2}(s^2/4)$$

where B denotes the beta function and $\mathcal{I}$ is the $_0F_1$ function from Chapters 3 and 4.

The Bessel function K_ν called *MacDonald's function* comes via the integral

$$\int_0^\infty \frac{\cos xy}{(y^2 + z^2)^{\nu+1/2}} \, dy = \left(\tfrac{1}{2}\right)_\nu^{-1} \left(\frac{x}{2z}\right)^\nu K_\nu(xz) \qquad (9.2.2)$$

for $\mathrm{Re}\,\nu > -1/2$, $x > 0$, $|\arg z| < \pi/2$. The function K_ν, for nonintegral ν can be expressed in terms of the Bessel function I_ν:

$$K_\nu(z) = \frac{\pi/2}{\sin \nu\pi} \left(I_{-\nu}(z) - I_\nu(z)\right)$$

(See Abramowitz&Stegun [1], Lebedev[32])

6. Verify the general approach to the binomial processes, using equation (9.2.1)

$$\int e^{-i\xi v} e^{tM(iv)} \frac{dv}{2\pi} = e^{\alpha\xi/\beta} \beta^{t/2} \int e^{-i\xi v} \left(-v^2 - \frac{\delta^2}{\beta^2}\right)^{t/2} \frac{dv}{2\pi}$$

where we must take care according to the signs of β and δ^2.

7. For the continuous binomial process, from Proposition 4.2.1, we have, via Problem 4, with $\beta = \sec^2\vartheta$, $\delta^2 = -1$:

$$e^{\alpha\xi/\beta} \beta^{t/2} \int e^{-i\xi v} \left(\frac{1}{\beta^2} - v^2\right)^{t/2} \frac{dv}{2\pi}$$

We see that the natural domain of integration is $|v| < \beta^{-1}$. Setting $v = u/\beta$, we have

$$e^{\alpha\xi/\beta} \beta^{-1-t/2} \int_{-1}^{1} e^{-i\xi u/\beta} \left(1 - u^2\right)^{t/2} \frac{du}{2\pi}$$

Thus, via Problem 5, the integral yields the factor

$$(2\pi)^{-1} B(1/2, 1 + t/2) \, \mathcal{I}_{(t+3)/2}(-\xi^2/4\beta^2)$$

8. For the negative binomial process, referring to Propositions 3.2.1 and 3.2.2, via Problem 4 we arrive at

$$\beta^{t/2} e^{\alpha\xi/\beta} \int e^{-i\xi v} \left(-v^2 - \frac{1}{4\beta^2}\right)^{t/2} \frac{dv}{2\pi}$$

Here we reverse the time as for the exponential process, setting $t = -2N$ Via equation (9.2.2), we find the result

$$\frac{(-1)^N}{\Gamma(N)} \frac{e^{\alpha\xi/\beta}}{\sqrt{\pi\beta}} \xi^{N-1/2} K_{N-1/2}(\xi/2\beta)$$

9. For the binomial process, referring to Propositions 2.2.1 and 2.2.2, we find the integrand to involve

$$\left[|\beta| \left(v^2 + \frac{1}{\beta^2} \right) \right]^{t/2}$$

Setting $t = -N$, the integral may be expressed in terms of MacDonald's function as in Problem 8.

9.3 SYMMETRIC FUNCTIONS AND KRAWTCHOUK POLYNOMIALS

There is an interesting interpretation of Krawtchouk polynomials in terms of symmetric functions, Ch. 4, §5.3.

1. Let $\{x_1, \ldots, x_N\}$ consist of k ones and $N - k$ negative ones. Show that the generating function for the elementary symmetric functions is

$$G(v, k) = (1 + v)^k (1 - v)^{N-k} = \sum_n v^n a_n$$

2. Interpreting the $+1$'s and -1's as steps of a random walk, let $x =$ position after N steps, starting from the origin. Compare with §2.5, and show that you recover the generating function G_t.

3. Express a_n in terms of polynomials K_n with $p = 1/2$.

4. By calculating $\langle G(v, X)G(v', X) \rangle$ where X has a binomial distribution,

   $\text{Prob}(X = k) = 2^{-N} \binom{N}{k}$, $0 \le k \le N$, show directly orthogonality of the a_n.

Chapter 6   BERNOULLI SYSTEMS

In this Chapter we study the analytic structure of Bernoulli systems. Although, in comparison with Chapter 5, the group theory and probability theory will be more in the background, they will still appear regularly. In the analytic vein, important features are: the Rodrigues formula (§1.3), the kernel ω (§II) which intertwines the Fock space realizations of Chapters 3 and 5, and the Riccati equation (§§1.4, V). Two integral transforms, the Fourier-Laplace transform and the gamma transform (§4.1) are basic elements of the construction of Bernoulli systems from the point of view of the present chapter.

The approach of this chapter is formulated so as to extend virtually verbatim to vector-valued or operator-valued processes. Thus, in many places it has an abstract flavor. With this analytic structure, one can look at higher-dimensional Lie algebras and corresponding groups and study representations that give Bernoulli systems in several variables. Details for some of the systems arising in this way will appear in Volume 3.

I.   Expansions, Rodrigues, and Riccati

We will look at some general expansions associated to GMS. For canonical GMS, we will find a Rodrigues formula for the Bernoulli systems. Then, connections with Riccati equations will be seen.

Remark. As usual when discussing expansions, a 'general function' denotes a polynomial or a linear combination of ('local') exponentials with polynomial coefficients.

1.1   EXPANSION THEOREMS FOR GENERALIZED MOMENT SYSTEMS

First let us consider a GMS in the variables (D, x).

1.1.1 Proposition.   Let $h_n(x, -t) = \exp(-tH) x^n$ be the time-reversed moment polynomials for the GMS with generator $H(z)$. Then we have the expansion formula:

$$f(x + y) = \sum_{n=0}^{\infty} \frac{h_n(y, -t)}{n!} e^{tH} D^n f(x)$$

Proof: From the generating function for the h_n, we have the expansion of the exponential function

$$e^{ax} = e^{tH(a)} \sum_{n=0}^{\infty} \frac{a^n}{n!} h_n(x, -t)$$

Replace $x \to y$, $a \to D$, and apply to $f(x)$. The result follows.   ∎

1.1.2 Corollary. *If* $\exp(tH(z))$ *is the moment generating function for the probability measure* p_t, *then*

$$f(x) = \sum_{n=0}^{\infty} \frac{h_n(x, -t)}{n!} \langle D^n f \rangle$$

where the expectation is with respect to p_t.

Proof: We have

$$e^{tH} f(0) = \int_{-\infty}^{\infty} f(y) \, p_t(dy) = \langle f \rangle$$

Apply this for $D^n f$ in the Proposition with $x \to 0$, $y \to x$.   ∎

Now consider a canonical GMS with generator $H(z)$ and canonical variable $V(z)$. Under the time-reversed flow, the time-zero polynomials $\xi_n(x)$ become $\xi_n(x, -t)$.

1.1.3 Proposition.   *For a canonical GMS we have the expansion formula*

$$f(x + y) = \sum_{n=0}^{\infty} \frac{\xi_n(y, -t)}{n!} e^{tH} V(D)^n f(x)$$

Proof: Start with the generating function, Ch. 5, Prop. 1.1.3,

$$e^{xU(v) - tM(v)} = \sum_{n=0}^{\infty} \frac{v^n}{n!} \xi_n(x, -t) \tag{1.1.1}$$

Now put $x \to y$, $v \to V(D)$ and apply to $f(x)$.   ∎

And, as in the above Corollary,

1.1.4 Corollary.   *For a canonical GMS, with* $\exp(tH(z))$ *the moment generating function for* p_t,

$$f(x) = \sum_{n=0}^{\infty} \frac{\xi_n(x, -t)}{n!} \langle V^n f \rangle$$

Remark. In this chapter, we work generally with a single probability measure, i.e., for a process, think of t as fixed. Thus, we drop the t subscript from p_t. Later on, t will be re-introduced, as a scaling factor in the parameters.

1.2 ORTHOGONALITY AND FOURIER-LAPLACE TRANSFORM

We have a probability measure $p \in \mathcal{P}^*$, with moment generating function $e^{H(z)}$. We consider an associated canonical GMS with canonical variable $V(z)$. Write

$$e^{-H(D)} \xi_n(x) = J_n(x)$$

We further assume the normalization $V'(0) = 1$, so that the $J_n(x)$ are monic (Ch. 4, Prop. 4.1.3). Now, suppose that the $\{ J_n \}$ are orthogonal polynomials. With squared norms $\gamma_n = \|J_n\|^2$, set

$$\phi_n(x) = (n!/\gamma_n) \, J_n(x)$$

Now we have

1.2.1 Theorem. *The Fourier-Laplace transform of $\phi_n(x) \, p(dx)$ is $V(z)^n e^{H(z)}$, i.e.,*

$$\int_{-\infty}^{\infty} e^{zx} \, \phi_n(x) \, p(dx) = V(z)^n e^{H(z)}$$

Thus, in Fourier space the polynomials $\phi_n(x)$ are realized via powers of the canonical variable V.

Proof: Substituting $v = V(z)$ in eq. (1.1.1) above, we have the expansion,

$$e^{zx} = e^{H(z)} \sum_{n=0}^{\infty} \frac{V(z)^n}{n!} \, J_n(x)$$

Now multiply by $\phi_n(x) \, p(dx)$ and integrate. By orthogonality, only one term survives, yielding

$$\int_{-\infty}^{\infty} e^{zx} \, \phi_n(x) \, p(dx) = e^{H(z)} \, V(z)^n (\gamma_n/n!) \, (n!/\gamma_n)$$

which gives the result.   ■

Remark. An interesting class of probability measures is constructed as follows. For $z \in \mathbf{R}$, normalize $e^{zx} p(dx)$ by dividing by $e^{H(z)}$, i.e., let

$$\mu_z(dx) = e^{zx - H(z)} \, p(dx)$$

Then $\mu_z \in \mathcal{P}^*$ for z in the interior of the domain around 0 where $H(z)$ is analytic. Denoting expectation with respect to μ_z by $\langle \; \rangle_z$, we have

$$\langle \phi_n \rangle_z = V(z)^n$$

exhibiting clearly the sense in which these polynomials may be thought of as powers. This construction arises in statistics, where the family μ_z is called an *exponential family* . (See Letac[34], Barndorff-Nielsen[3] for more on exponential families.)

Now let us take $n = 1$ in Theorem 1.2.1. Then we have

$$\int_{-\infty}^{\infty} e^{zx} \, \phi_1(x) \, p(dx) = V(z) e^{H(z)}$$

with ϕ_1 a polynomial of degree 1. Differentiating $\int e^{zx} p(dx) = e^{H(z)}$ with respect to z, we have

$$\int_{-\infty}^{\infty} e^{zx} \, x \, p(dx) = H'(z) e^{H(z)}$$

and, recalling Ch. 4, Prop. 1.2.2, with the normalization $V'(0) = 1$, we have

$$H'(z) = \mu + \sigma^2 V(z) \tag{1.2.1}$$

with μ and σ^2 the mean and variance of the distribution p. And we see that $\phi_1(x) = (x - \mu)/\sigma^2$.

1.3 RODRIGUES FORMULA

A *Rodrigues formula* for a family of orthogonal polynomials, $\{\phi_n\}$, say, with respect to the measure p, is a formula of the type $\phi_n = p^{-1} Q^n p$, where Q is a particular operator, e.g., a differential operator. Usually some additional factors, generally functions of x, are involved as well. We will see that for the Bernoulli systems the Rodrigues formula has a simple form, directly related to the result of Theorem 1.2.1. First we must discuss what we mean by the action of an operator of the form $\varphi^*(D)$ on $p(dx)$, where φ^* denotes the formal adjoint of φ. We define $\varphi^*(D)p(dx)$ via Fourier-Laplace transform:

$$\int_{-\infty}^{\infty} e^{zx} \, \varphi^*(D) p(dx) = \int_{-\infty}^{\infty} (\varphi(D)e^{zx}) \, p(dx) = \varphi(z) \int_{-\infty}^{\infty} e^{zx} \, p(dx) \tag{1.3.1}$$

When p has a density, i.e., $p(dx) = p(x) \, dx$, then we have the usual interpretations of the action on the function $p(x)$.

1.3.1 Theorem.   *The polynomials $\{\phi_n\}$ are given by the Rodrigues formula*

$$\phi_n(x) = p(dx)^{-1} V^*(D)^n p(dx)$$

where $V^*(z) = V(-z)$ is the formal adjoint of V (for integration by parts with respect to dx).

Proof: Take the Fourier-Laplace transform of the Rodrigues formula:

$$\int_{-\infty}^{\infty} e^{zx}\,(p(dx)^{-1}V^*(D)^n p(dx))\,p(dx) = \int_{-\infty}^{\infty} e^{zx}\,V^*(D)^n p(dx)$$

$$= V(z)^n \int_{-\infty}^{\infty} e^{zx}\,p(dx)$$

as in eq. (1.3.1). Comparing with Theorem 1.2.1 yields the result. ∎

Remark. With $V(0) = 0$, V acts nilpotently on ϕ_n, so that the Rodrigues formula readily implies orthogonality of the ϕ_n. E.g., with $m > n$,

$$\int_{-\infty}^{\infty} \phi_m(x)\phi_n(x)\,p(dx) = \int_{-\infty}^{\infty} \phi_n(x)V^*(D)^m p(dx) = \int_{-\infty}^{\infty} (V^m \phi_n(x))\,p(dx) = 0$$

1.4 RICCATI EQUATION AND ORTHOGONALITY

Beginning with the formula from Theorem 1.2.1,

$$\int_{-\infty}^{\infty} e^{zx}\,\phi_n(x)\,p(dx) = V(z)^n\,e^{H(z)}$$

differentiate with respect to z. This gives

$$\int_{-\infty}^{\infty} e^{zx}\,(x\phi_n(x))\,p(dx) = (nV(z)^{n-1}V'(z) + V(z)^n H'(z))\,e^{H(z)} \tag{1.4.1}$$

$$= (nV(z)^{n-1}V'(z) + (\mu + \sigma^2 V(z))V(z)^n)\,e^{H(z)}$$

using eq. 1.2.1.

A three-term recurrence relation for the polynomials ϕ_n, say of the form

$$x\phi_n = \kappa_n\phi_{n+1} + \alpha_n\phi_n + \beta_n\phi_{n-1}$$

thus corresponds to V' expressed as a function of V. This shows the significance of the Riccati equation, namely, the three terms of the recurrence correspond to the quadratic polynomial $\pi(v)$, where we have $V' = \pi(V)$.

1.4.1 Proposition. *Let V satisfy the Riccati equation $V' = \pi(V)$, with $\pi(v) = 1 + 2\alpha v + \beta v^2$, the characteristic polynomial. Then we have the recurrence*

$$x\phi_n = (\sigma^2 + \beta n)\phi_{n+1} + (\mu + 2\alpha n)\phi_n + n\phi_{n-1}$$

with $\phi_0 = 1$, $\phi_1 = (x - \mu)/\sigma^2$.

   Proof: Substituting the Riccati equation into the right side of eq. (1.4.1) yields, dropping the factor $e^{H(z)}$,

$$nV^{n-1}V' + (\mu + \sigma^2 V)V^n = nV^{n-1} + (\mu + 2\alpha n)V^n + (\sigma^2 + \beta n)V^{n+1}$$

Now we use the correspondence, Theorem 1.2.1, between powers of V and the polynomials ϕ_n to read off the recurrence relation.   ∎

Remark. Notice that the converse holds as well, namely, that the recurrence relation for the ϕ_n implies the Riccati equation, again via the Fourier-Laplace correspondence.

And for the monic polynomials $\{\,J_n\,\}$ we have

1.4.2 Proposition.   *The monic polynomials $\{\,J_n\,\}$ are given by*

$$J_n(x) = \beta^n (\sigma^2/\beta)_n\, \phi_n(x)$$

They satisfy the recurrence

$$xJ_n = J_{n+1} + (\mu + 2\alpha n)J_n + n(\sigma^2 + \beta(n-1))J_{n-1}$$

with $J_0 = 1$, $J_1 = x - \mu$, and squared norms $\gamma_n = n!\,\beta^n(\sigma^2/\beta)_n$.

   Proof: Substitute in the above Proposition as indicated .   ∎

At this point, we can work with $\beta = 0$ as well as $\beta \neq 0$, corresponding to HW and sl(2) Fock spaces respectively. Now we introduce t as a scale factor, implicitly taking $\beta \neq 0$ as 'generic'. This corresponds in our study of limit theorems in Chapter 5 to the scaling $t \to t/\beta$, $\beta \to 0$ to recover the HW Fock space and HW and oscillator algebras. We have

1.4.3 Corollary.   *Determine s and t by the relations*

$$\mu = \alpha s, \qquad \sigma^2 = \beta t$$

then the recurrence for $\{\,J_n\,\}$ is

$$xJ_n = J_{n+1} + \alpha(s + 2n)J_n + \beta n(t + n - 1)J_{n-1}$$

with $J_0 = 1$, $J_1 = x - \alpha s$, and squared norms

$$\gamma_n = n!\,\beta^n(t)_n$$

And we state the form we will use for the ϕ_n

1.4.4 Proposition.   *The ϕ_n satisfy the recurrence*

$$x\phi_n = \beta(t + n)\phi_{n+1} + \alpha(s + 2n)\phi_n + n\phi_{n-1}$$

with $\phi_0 = 1$, $\phi_1 = (x - \alpha s)/\beta t$, and squared norms

$$\|\phi_n\|^2 = n!/\beta^n(t)_n$$

We are ready for our study of the kernel ω.

II. ω-kernel

We introduce the generating function for the ϕ_n. We will see that its properties tell us much about the structure of the Bernoulli systems.

2.1 Definition.   The ω-kernel is the generating function

$$\omega(y, x) = \sum_{n=0}^{\infty} \frac{y^n}{n!} \phi_n(x)$$

Remark. We state immediately that we have as well

$$\omega(y, x) = \sum_{n=0}^{\infty} \frac{y^n J_n(x)}{\gamma_n} \tag{2.1}$$

via the relations $\phi_n = (n!/\gamma_n) J_n$.

2.2 Theorem.   *The ω-kernel is an intertwining kernel between two realizations of the sl(2) Fock space. Namely,*

1. *For the inner product given by integration with respect to the measure p we have*

$$\int_{-\infty}^{\infty} \omega(y, x) J_n(x) p(dx) = y^n$$

2. *For the inner product on the Fock space with orthogonal basis y^n, with $\|y^n\|^2 = \gamma_n = n! \beta^n(t)_n$, we have*

$$\langle \omega(y, x), y^n \rangle = J_n(x)$$

Proof: Follows directly from eq. (2.1). Note that in the last relation the 'variable-of-integration' in the Fock space inner product is y.   ■

2.1 ω AND RODRIGUES

We compute the transform of ω

2.1.1 Proposition.   *The Fourier-Laplace transform of ω:*

$$\int_{-\infty}^{\infty} e^{zx} \omega(y, x) p(dx) = e^{y V(z) + H(z)}$$

Proof: Start with the basic relation, Theorem 1.2.1,

$$\int_{-\infty}^{\infty} e^{zx}\, \phi_n(x)\, p(dx) = V(z)^n e^{H(z)}$$

Multiplying by $y^n/n!$ and summing over n yields the result. ∎

Recalling the discussion of the Rodrigues formula, §1.3, we see that

2.1.2 Theorem. *The ω-kernel can be expressed as a generating function version of the Rodrigues formula:*

$$\omega(y,x) = p(dx)^{-1} e^{y V^*(D)}\, p(dx)$$

III. X operator

Using the kernel ω we can formulate the operator X, given as multiplication by x in the realization acting on the ϕ_n, as a differential operator acting on functions of y.

3.1 Theorem. *The operator X acting on functions of y, i.e., the sl(2) Fock space with basis $\{\, y^n \,\}$, is*

$$X = \alpha s + \beta t \frac{d}{dy} + y\,\pi\left(\frac{d}{dy}\right)$$

with $\pi(v) = 1 + 2\alpha v + \beta v^2$. As a function of y for given x, we have

$$X\omega(y,x) = x\,\omega(y,x)$$

i.e., $\omega(y,x)$ are eigenfunctions of X with corresponding eigenvalues x.

Proof: Via the recurrence, Prop. 1.4.4, we have

$$x\omega(y,x) = \sum_{n=0}^{\infty} \frac{y^n}{n!} \left(\beta(t+n)\phi_{n+1} + \alpha(s+2n)\phi_n + n\phi_{n-1}\right)$$

$$= \sum_{n=0}^{\infty} \frac{(Xy^n)}{n!}$$

On ω we have the correspondence:

$$\phi_{n+1} \leftrightarrow d/dy, \qquad n\phi_{n-1} \leftrightarrow y, \qquad n\phi_n \leftrightarrow y\, d/dy$$

Translating the recurrence via this correspondence gives the result. ∎

This shows precisely how the spectrum of the operator X is the support of the measure p, i.e., points such that every neighborhood has non-zero probability. Thus, the Bernoulli systems provide specific realizations of the operator algebra formulation of probability theory.

3.1 ANALYTIC FORM OF ω

As we have just seen, ω can be found by solving for the eigenfunctions of X, i.e., let us solve $Xu = xu$, for u regular at the origin, $u(0) = 1$. When we refer to 'the' solution of the differential equation, even though it is of second order, we implicitly mean the solution satisfying the regularity conditions just mentioned.

3.1.1 Lemma.   *The solution $u(y)$ to the confluent hypergeometric equation*

$$yu'' + (By + C)u' = Eu$$

where B, C, E are constants, is

$$u(y) = {}_1F_1\left(\begin{matrix} -E/B \\ C \end{matrix} \;\middle|\; -By\right)$$

Proof: The result may be checked directly. However, it is interesting here to recall the number operator for the Laguerre polynomial system, Ch. 5, Prop. 6.4.1. With $RV = xz(1 - \alpha z) - \alpha tz$, we have $u = L_n(x, t)$ satisfying

$$(x - \alpha t)u' - \alpha x u'' = nu$$

Rewrite this in the form

$$xu'' + (-(x/\alpha) + t)u' = (-n/\alpha)u$$

From the form of L_n as a hypergeometric function, Ch. 5, Prop. 6.4.2, we see that the solution we seek is given by

$$u = {}_1F_1\left(\begin{matrix} -n \\ t \end{matrix} \;\middle|\; x/\alpha\right)$$

The only question is whether the fact that n is an integer affects the generality of the solution. One can check readily that it does not.   ∎

3.1.2 Theorem.   *The kernel ω is given by*

$$\omega(y, x) = e^{y/(\delta - \alpha)} \, {}_1F_1\left(\begin{matrix} (x + \delta t + \alpha(t - s))/2\delta \\ t \end{matrix} \;\middle|\; \frac{2\delta}{\beta} y\right)$$

Proof: From Theorem 3.1, we have the equation $Xu = xu$:

$$\beta y u'' + (2\alpha y + \beta t)u' + (y + \alpha s)u = xu \tag{3.1.1}$$

First we eliminate the 'yu' term. Let $u(y) = e^{Ay} w(y)$. Substituting into eq. (3.1.1), one finds the coefficient of yw to be $\pi(A)$, the characteristic polynomial. Recall the factorization (as in the proof of Ch. 5, Theorem 1.5.2.1)

$$\pi(A) = (1 + (\alpha + \delta)A)(1 + (\alpha - \delta)A)$$

Take the root $A = 1/(\delta - \alpha)$. The resulting equation for w is

$$\beta y w'' + (\beta t - 2\delta y)w' = (x + \delta t + \alpha(t - s))w$$

where we use the relation $\beta/(\alpha - \delta) = \alpha + \delta$. After dividing through by β, the result follows from Lemma 3.1.1. ∎

Now we can find the form of the polynomials ϕ_n.

3.1.3 Lemma. *We have the generating function for $_2F_1$ polynomials:*

$$e^a \, _1F_1 \left(\left. \begin{matrix} b \\ c \end{matrix} \right| x \right) = \sum_{n=0}^{\infty} \frac{a^n}{n!} \, _2F_1 \left(\left. \begin{matrix} -n, b \\ c \end{matrix} \right| -\frac{x}{a} \right)$$

Proof: This follows by writing out the $_2F_1$ and summing. ∎

And for the ϕ_n

3.1.4 Theorem. *The polynomials ϕ_n are given by*

$$\phi_n(x) = (\delta - \alpha)^{-n} \, _2F_1 \left(\left. \begin{matrix} -n, (x + \delta t + \alpha(t - s))/2\delta \\ t \end{matrix} \right| \frac{2\delta}{\alpha + \delta} \right)$$

Proof: This follows from Theorem 3.1.2 and the above Lemma. ∎

Remark. We can check that this agrees with the polynomials J_n that we have from Chapter 5. With $\phi_n = (n!/\gamma_n)J_n$, see also Prop. 1.4.2 and the Corollary there, we have $J_n = \beta^n(t)_n \phi_n$. And we find the result, here with t as an implicit argument,

$$J_n(x) = (-1)^n (t)_n (\alpha + \delta)^n \, _2F_1 \left(\left. \begin{matrix} -n, (x + \delta t + \alpha(t - s))/2\delta \\ t \end{matrix} \right| \frac{2\delta}{\alpha + \delta} \right)$$

via $\beta/(\alpha - \delta) = \alpha + \delta$. The difference is that in Chapter 5 we have a convolution semigroup of measures. Recall that s and t are defined as proportionality factors $\mu = \alpha s$, $\sigma^2 = \beta t$. And for the measures p_t, we have $\mu = \alpha t$, $\sigma^2 = \beta t$, i.e., $s = t$, and the additional drift term drops out. Cf., Corollary to Theorem 1.5.1.1 of Chapter 5.

3.2 MOMENT POLYNOMIALS

Here we have a family of moment polynomials with raising operator X.

3.2.1 Definition.   The moment polynomials $m_n(y)$ are defined by

$$m_n(y) = X^n 1$$

We will find an integral formula for the polynomials m_n. First we look at the generating function.

3.2.2 Proposition.   Let $g(z,y) = e^{zX} 1$ be the generating function of the polynomials $m_n(y)$. Then:

1. $g(z,y)$ is the solution to

$$\frac{\partial u}{\partial z} = Xu, \qquad u(0,y) = 1$$

2. The integral formula

$$g(z,y) = \int_{-\infty}^{\infty} e^{zx}\, \omega(y,x)\, p(dx)$$

3. The exponential formula

$$g(z,y) = e^{yV(z)+H(z)}$$

Proof: Statement #1 is by construction. For #2, by orthogonality of the ϕ_n, we have

$$\int_{-\infty}^{\infty} \omega(y,x)\, p(dx) = \phi_0 = 1, \qquad \forall y$$

Since $\omega(y,x)$ are eigenfunctions of X with eigenvalues x, applying $\exp(zX)$ gives

$$g(z,y) = e^{zX} 1 = \int_{-\infty}^{\infty} e^{zx}\, \omega(y,x)\, p(dx)$$

I.e., $g(z,y)$ is the Fourier-Laplace transform of ω. And #3 follows by Prop. 2.1.1. ∎

Now we can describe the m_n.

3.2.3 Proposition.   The polynomials m_n have the following properties:

1. Integral formula:

$$m_n(y) = \int_{-\infty}^{\infty} x^n \omega(y,x)\, p(dx)$$

2. The values $m_n(0)$ are the moments of p:

$$m_n(0) = \mu_n = \int_{-\infty}^{\infty} x^n\, p(dx)$$

Proof: The first statement follows directly from the above Proposition. And then, setting $y = 0$, #2 follows via the observation that $\omega(0, x) = \phi_0 = 1$ for all x. ∎

Let us look at the generating function. Differentiate the relation #3 in Prop. 3.2.2 with respect to z:

$$X\, g(z, y) = (yV'(z) + H'(z))\, e^{yV(z)+H(z)}$$
$$= (yV'(z) + \alpha s + \beta t V(z))\, e^{yV(z)+H(z)}$$
$$= (y\pi(V(z)) + \alpha s + \beta t V(z))\, e^{yV(z)+H(z)}$$

via eq. 1.2.1 (with appropriate substitutions made for μ and σ^2) and the Riccati equation for V. As seen from the exponential form, on $g(z, y)$ we have the correspondence $V \leftrightarrow d/dy$. Thus, we recover the form of X as a differential operator as in Theorem 3.1.

IV. Generating functions

Now we will see the connection between the ω-kernel, the generating function $G(v, x) = \sum(v^n/n!)J_n(x)$, and the generating function for the scaled squared norms $\gamma_n/n!$. Here $G(v, x)$ is the generating function $G_t(v, x)$ of Chapter 5, but we are dropping explicit t-dependence, so that we have

$$G(v, x) = \exp(xU(v) - M(v)) \qquad (4.1)$$

where, as usual, U is the function inverse to V and M is the canonical generator, $M(v) = H(U(v))$.

4.1 GAMMA TRANSFORM

Here we will use another integral transform. We use a gamma integral:

$$\int_0^\infty x^{n+t-1} e^{-x}\, dx/\Gamma(t) = (t)_n$$

Now we have the

4.1.1 Definition. The *gamma transform* of a function f is given by

$$\Gamma_t f(v) = \int_0^\infty f(xv)\, x^{t-1} e^{-x}\, dx/\Gamma(t)$$

Remark. Note that this is expressed in terms of expectation with respect to the gamma distribution (Ch. 5, Def. 6.1) with density $p_t(x) = x^{t-1} e^{-x}/\Gamma(t)$. If X_t is the corresponding random variable, we have

$$\Gamma_t f(v) = \langle f(vX_t)\rangle$$

The following useful formulas are immediate. Note that the second transform is the moment generating function for the gamma distribution, with v as a scale factor.

4.1.2 Lemma.   *Some gamma transforms:*

1. *Let $f_n(x) = x^n$ denote the power function. Then*

$$\Gamma_t f_n(v) = v^n(t)_n = (t)_n f_n(v)$$

2. *For the exponential function $e_s(x) = e^{sx}$ we have*

$$\Gamma_t e_s(v) = (1 - sv)^{-t}$$

4.2 GENERATING FUNCTIONS

Now we will see how the various generating functions are related.

4.2.1 Theorem.   *The gamma transform of $\omega(y, x)$, suitably scaled, is $G(v, x)$:*

$$\Gamma_t \omega(\beta v, x) = G(v, x)$$

Proof: Using #1 of Lemma 4.1.2:

$$\Gamma_t \omega(\beta v, x) = \int_0^\infty e^{-y}\, \omega(\beta v y, x)\, y^{t-1} dy / \Gamma(t)$$

$$= \sum_{n=0}^\infty \frac{v^n}{n!}\, \beta^n (t)_n \phi_n(x)$$

$$= \sum_{n=0}^\infty \frac{v^n}{n!}\, J_n(x) = G(v, x)$$

■

Next, we have

4.2.2 Theorem.   *The Fourier-Laplace transform of $G(v, x)$:*

$$\int_{-\infty}^\infty e^{zx}\, G(v, x)\, p(dx) = (1 - \beta v V(z))^{-t} e^{H(z)}$$

Proof: Start with the Fourier-Laplace transform of ω, Prop. 2.1.1:

$$\int_{-\infty}^\infty e^{zx}\, \omega(y, x)\, p(dx) = e^{y V(z) + H(z)}$$

Now apply the gamma transform, with notation as in Lemma 4.1.2:

$$\int_{-\infty}^\infty e^{zx}\, \Gamma_t \omega(\beta v, x)\, p(dx) = \Gamma_t e_{\beta V(z)}(v)\, e^{H(z)}$$

Now, for the left side apply Theorem 4.2.1, and for the right side apply #2 of Lemma 4.1.2. The result follows.   ■

This result will be discussed further in the next section. Here we continue to derive one aspect of the significance of this result.

4.2.3 Definition.   The function $\phi(x)$ is given by

$$\phi(x) = -t \log(1 - \beta x)$$

The above Theorem may be formulated as

4.2.4 Proposition.   *The Fourier-Laplace transform of* $G(v, x)$ *is* $e^{H(z)+\phi(vV(z))}$.

This next relation is called a *cocycle relation* for the generator H. It shows that ϕ measures the deviation from linearity of H. Recall that H is linear only in the simplest case of the Gaussian distribution (Ch. 5, §VII).

4.2.5 Theorem.   *The generator H satisfies the cocycle relation*

$$H(a+b) - H(a) - H(b) = \phi(V(a)V(b))$$

Proof: Substitute eq. (4.1) in Theorem 4.2.2:

$$\int_{-\infty}^{\infty} e^{zx}\, e^{xU(v)-M(v)}\, p(dx) = e^{H(z)+\phi(vV(z))}$$

On the other hand, direct calculation, using the fact that $\exp(H(z))$ is the moment generating function of $p(dx)$, yields

$$\int_{-\infty}^{\infty} e^{zx}\, e^{xU(v)-M(v)}\, p(dx) = e^{H(z+U(v))-H(U(v))}$$

Equating the right hand sides, with the substitutions $z \to a$, $v \to V(b)$, rearranging, and taking logarithms, the result follows.   ■

Remark. Here we see that it is quite natural to consider the convolution semigroup p_t, so that H is replaced by tH. Then in the definition of ϕ, there is no factor of t involved.

4.3   ORTHOGONALITY AND THE FUNCTION ϕ

Now we can see the main significance of the function ϕ.

4.3.1 Theorem.   *A cocycle relation of the form*

$$H(a+b) - H(a) - H(b) = \phi(V(a)V(b))$$

for a generator H implies that the corresponding canonical GMS is a family of orthogonal polynomials. The generating function for the scaled squared norms $\gamma_n/n!$ is given by:

$$\sum_{n=0}^{\infty} \frac{a^n}{n!}\,(\gamma_n/n!) = e^{\phi(a)} \tag{4.3.1}$$

Proof: Start with the generating function

$$G(v, x) = e^{xU(v)-M(v)} = \sum_{n=0}^{\infty} \frac{V(z)^n}{n!} J_n(x)$$

which makes no *a priori* assumptions about orthogonality. Then we have, similar to the proof of Theorem 4.2.5:

$$\int_{-\infty}^{\infty} G(V(a), x) G(V(b), x) \, p(dx) = e^{H(a+b)-H(a)-H(b)}$$

On the other hand, the left side equals, where the indicated expectation is with respect to $p(dx)$:

$$\sum_{m=0}^{\infty}\sum_{n=0}^{\infty} \frac{V(a)^m V(b)^n}{m!\,n!} \langle J_m(X) J_n(X) \rangle$$

Thus, this is a function only of the product $V(a)V(b)$ if and only if the J_n are orthogonal. In the case of orthogonality, with γ_n denoting the squared norms, using the cocycle relation, we have, equating right hand sides:

$$e^{\phi(V(a)V(b))} = \sum_{n=0}^{\infty} \frac{(V(a)V(b))^n}{n!\,n!} \gamma_n \qquad (4.3.2)$$

Replacing $V(a)V(b)$ by a, the result follows.   ■

Notice that the knowledge of ϕ is thus equivalent to knowing the squared norms $\{\gamma_n\}$. And we see easily that Def. 4.2.3 is in agreement with the values $\gamma_n = n!\,\beta^n(t)_n$.

Remark. For the Poisson and Gaussian systems, we have $\beta = 0$. From the squared norms $\gamma_n = n!\,t^n$, we see, via eq. (4.3.1), that we should have $\phi(x) = tx$. With the explicit scaling $H \to tH$ we have

$$\phi(x) = x \qquad (4.3.3)$$

Notice from Def. 4.2.3 that this is consistent with the Poisson and diffusion limits, $t \to t/\beta$, $\beta \to 0$, of Chapter 5.

V.   Riccati equations and Bernoulli systems

Now we will use group theory in connection with solving the Riccati equations that are fundamental to Bernoulli systems. The main feature is that:

fractional linear transformations map Riccati equations into Riccati equations

We will clarify the meaning of this statement and see another way in which groups provide the structure of the Bernoulli systems.

5.1 FRACTIONAL LINEAR TRANSFORMATIONS

First we recall the definition of fractional linear transformation, which we abbreviate to *FLT* .

5.1.1 Definition. A *fractional linear transformation* is a mapping of the form:

$$x \rightarrow \frac{ax + b}{cx + d}$$

where a, b, c, d are constant parameters.

For reference and to get some feeling for the mappings, we have

5.1.2 Proposition. *Let* $T(x) = (ax + b)/(cx + d)$ *be a FLT. Then*

1. *The derivative*

$$T'(x) = \frac{ad - bc}{(cx + d)^2}$$

2. *The inverse mapping, for* $ad - bc \neq 0$,

$$T^{-1}(x) = \frac{dx - b}{-cx + a}$$

We see that is natural to identify each FLT with the corresponding matrix, $\begin{pmatrix} a & b \\ c & d \end{pmatrix}$, say. Thus, the set of FLT's with non-zero determinant, $ad - bc$, form a group. What is important here is to notice that any common scale factor drops out. Thus, once the determinant is non-zero, we may take it to be scaled to 1.

5.1.3 Definition. The group *PSL(2)* is the group of FLT's with determinant 1.

Equivalently, it is the group SL(2) of 2×2 matrices of determinant 1 modulo the equivalence relation $A_1 \sim A_2$ if there exists a non-zero scalar λ such that $A_1 = \lambda A_2$. Under this equivalence we have the *projective special linear group* thus denoted PSL(2).

Another way of looking at this group is to consider the linear transformations on projective space.

5.1.4 Definition. Given a vector space $\mathcal{V}$, the corresponding *projective space* $P\mathcal{V}$, is the quotient of $\mathcal{V}$ by the equivalence relation

$$v_1 \sim v_2 \qquad \Leftrightarrow \qquad \exists \lambda \neq 0, \quad v_1 = \lambda v_2$$

With $\mathcal{V}$ as $\mathbf{R}^2$ or $\mathbf{C}^2$, we identify the pair (x, y), for $y \neq 0$ with the pair $(x/y, 1)$. We see that all pairs $(x, 0)$, with $x \neq 0$, are equivalent, giving the point at infinity. So if you make a linear transformation:

$$\begin{pmatrix} X \\ Y \end{pmatrix} = \begin{pmatrix} ax + by \\ cx + dy \end{pmatrix}$$

for $Y, y \neq 0$, then under the equivalence this is the same as

$$X/Y = (ax + by)/(cx + dy)$$

And with the variables $U = X/Y$, $u = x/y$, we have the FLT: $U = (au + b)/(cu + d)$. Thus we see that

5.1.5 Proposition.   *The group PSL(2) may be identified with SL(2) of linear transformations with determinant 1 acting on the corresponding projective space.*

Now we are ready to see the connection with Riccati equations.

5.2 RICCATI EQUATIONS AND FLT'S

We will use the pair (Y, y) as a generic variable in projective space. Then the variable $V = Y/y$, so that linear transformations of the pair (Y, y) correspond to FLT's of V. The main feature is that as linear transformations correspond to FLT's, linear differential equations (here with constant coefficients) correspond to Riccati equations.

5.2.1 Proposition.   *Let $(Y(z), y(z))$ satisfy the linear differential equation*

$$\begin{pmatrix} Y \\ y \end{pmatrix}' = \begin{pmatrix} a & b \\ c & d \end{pmatrix} \begin{pmatrix} Y \\ y \end{pmatrix}, \qquad \begin{pmatrix} Y(0) \\ y(0) \end{pmatrix} = \begin{pmatrix} Y_0 \\ y_0 \end{pmatrix}$$

Then $V(z) = Y(z)/y(z)$ satisfies the Riccati equation

$$V' = b + (a - d)V - cV^2, \qquad V(0) = Y_0/y_0$$

Proof: Write out $Y' = aY + by$, $y' = cY + dy$. Then using $V' = (yY' - Yy')/y^2$, the result follows directly. ∎

Observe that there are *two* parameters in the initial condition for the linear system, whereas the Riccati equation has one initial condition, involving only the ratio Y_0/y_0. The Riccati equation determines $V'(0)$ from $V(0)$, which is determined by the initial values Y_0, y_0, and these as well determine the initial values $Y'(0), y'(0)$ via the differential equation. One easily checks that this is all consistent.

Looking at the coefficients, we see that only the difference $a - d$ appears. We will use later the form corresponding to sl(2). I.e., we take the matrix corresponding to the Riccati equation to have trace zero, $d = -a$. Then we have the correspondence

$$V' = b + 2aV - cV^2 \quad \leftrightarrow \quad \begin{pmatrix} a & b \\ c & -a \end{pmatrix} \tag{5.2.1}$$

We can immediately give the solutions. Write $A = \begin{pmatrix} a & b \\ c & d \end{pmatrix}$.

5.2.2 Proposition.   *The solution to the linear differential equation*

$$\begin{pmatrix} Y \\ y \end{pmatrix}' = A \begin{pmatrix} Y \\ y \end{pmatrix}, \qquad \begin{pmatrix} Y(0) \\ y(0) \end{pmatrix} = \begin{pmatrix} Y_0 \\ y_0 \end{pmatrix}$$

given by

$$\begin{pmatrix} Y(z) \\ y(z) \end{pmatrix} = \exp(zA) \begin{pmatrix} Y_0 \\ y_0 \end{pmatrix}$$

gives the solution $V(z) = Y(z)/y(z)$, $V(0) = Y_0/y_0$, to the Riccati equation.

And we have for $A \in \text{sl}(2)$,

5.2.3 Corollary.   *If the coefficient matrix A of the linear differential equation is in sl(2), the solution is given by the corresponding one-parameter subgroup of SL(2) generated by A.*

Now we want to see how solutions correspond under FLT's.

5.2.4 Proposition.   *Let $V = Y/y$ where $\binom{Y}{y}$ satisfies the equation $\binom{Y}{y}' = A\binom{Y}{y}$. Let $v = (\alpha V + \beta)/(\gamma V + \delta)$ be a FLT of V with coefficient matrix $E = \left(\begin{smallmatrix}\alpha & \beta \\ \gamma & \delta\end{smallmatrix}\right)$. Then, $v = Y/y$, where $\binom{Y}{y}$ satisfies the equation*

$$\binom{Y}{y}' = (EAE^{-1})\binom{Y}{y}$$

i.e., the corresponding coefficient matrices are related by a similarity transformation.

Proof: The equation with coefficient matrix EAE^{-1} may be rewritten

$$\left(E^{-1}\binom{Y}{y}\right)' = AE^{-1}\binom{Y}{y}$$

Thus, $E^{-1}\binom{Y}{y}$ corresponds to V. Hence, the corresponding $v = Y/y$ for the stated equation is the FLT of V corresponding to the matrix E.   ∎

Remark. Here we can see how the notion of a *gauge group* comes up. That is, if the matrices E and A commute, then the differential equation remains invariant. Thus, for given A, the set of nonsingular matrices commuting with A is a group, which may be thought of as a gauge group for the equation. Thus, the corresponding solutions are distinguished only by the initial conditions.

Now we see the basic feature of the Riccati systems. Namely, that we produce solutions from a given solution by FLT's. Since the corresponding coefficient matrices of the linear systems are related by similarity, the spectrum remains invariant under the group action. Equivalently, given A, we can take $\operatorname{tr} A$ and $\det A$ as invariants. If we further restrict to $\operatorname{tr} A = 0$, this leaves the determinant as the only invariant. Thus,

5.2.5 Proposition.   *Let $V' = b + 2aV - cV^2$. Let v be related to V by an FLT. Then v satisfies a Riccati equation $v' = B + 2Av - Cv^2$ such that $a^2 + bc = A^2 + BC$.*

Proof: The expression $a^2 + bc$ is just minus the determinant of the corresponding coefficient matrix of the linear system.   ∎

Now we are ready to study the Bernoulli systems.

5.3 FLT'S AND BERNOULLI SYSTEMS

For the Bernoulli systems, we have the canonical variable determined by the Riccati equation $V' = 1 + 2\alpha V + \beta V^2$, with $V(0) = 0$. We take the corresponding linear system to be

$$\begin{pmatrix} Y \\ y \end{pmatrix}' = \begin{pmatrix} \alpha & 1 \\ -\beta & -\alpha \end{pmatrix} \begin{pmatrix} Y \\ y \end{pmatrix}$$

with initial condition $\begin{pmatrix} 0 \\ 1 \end{pmatrix}$. Now we consider transformations of the Bernoulli systems induced by FLT's acting on V.

5.3.1 Proposition.   *For the Bernoulli systems, we have δ^2 as an invariant under the group of FLT's.*

Proof: As we have seen above, the determinant is invariant. Here we have the determinant $-\alpha^2 + \beta = -\delta^2$. I.e., we are taking $-$ det.   ∎

So let us classify the Bernoulli systems according to δ^2.

5.3.2 Proposition.   *The Bernoulli systems may be classified as follows:*

1. *For $\delta^2 > 0$, we have the binomial systems, binomial and negative binomial, and the Poisson system.*

2. *For $\delta^2 < 0$, we have the continuous binomial system.*

3. *For $\delta^2 = 0$, we have the exponential system and the Gaussian system.*

Recall that δ mainly has the interpretation as a scale for the increments of the process. For real δ we have the binomial and Poisson processes, while the continuous binomial corresponds to imaginary δ. While Brownian motion indeed has continuous paths, it is not the case for the exponential process.

A convenient way to see how the above classification works is to look at the generator

$$H(z) = \log \left(\frac{\delta \operatorname{sech} \delta z}{\delta - \alpha \tanh \delta z} \right)$$

We see that the sign of δ^2, i.e., whether δ is real or imaginary, gives, respectively, hyperbolic or trigonometric functions in the generator. When $\delta = 0$, we have rational functions. The Poisson process is special from this point of view. With $\beta = 0$, we have $\delta^2 = \alpha^2$, so that since the jump size α is real, $\delta \in \mathbf{R}$ as well, while the generator itself is already an exponential function.

So let us look at the corresponding orbits. We will break up into two classes $\delta \neq 0$ and $\delta = 0$. Over the reals, you cannot get the continuous binomial system from the Poisson, as $\alpha^2 = \delta^2 > 0$. But from the analytic point of view, you can generate it by using imaginary α. The set of Riccati equations, and hence systems, related by FLT's constitute an *orbit* of the group of FLT's acting on them. Thus, we consider two such orbits.

Now consider the conditions we have on V. We have $V(0) = 0$ and we normalize $V'(0) = 1$.

5.3.3 Proposition. *The elements of the group of FLT's acting on Bernoulli systems preserving the conditions $V(0) = 0$, $V'(0) = 1$, have the form*

$$\begin{pmatrix} 1 & 0 \\ c & 1 \end{pmatrix}$$

Thus, this group is isomorphic to the translation group.

Proof: With $v = (aV + b)/(cV + d)$, we have $V(0) = 0 \Rightarrow b = 0$. With $b = 0$, taking derivatives, we have

$$v' = \frac{ad}{(cV + d)^2} V'$$

Evaluating at 0 gives $1 = a/d$. Making use of the free scale factor, we can scale out $a = d$ to 1, leaving c as the remaining effective parameter. It is readily checked that matrices of the form stated provide a representation of the translation group, their product corresponding to addition of the parameters. ∎

5.3.1 Exponential-Gaussian orbit

Start with the simplest Riccati equation $V' = 1$, corresponding to the Brownian motion process (Gaussian system).

5.3.1.1 Proposition. *Under the action of the group of FLT's the Gaussian system maps to the exponential system. Under the transformation $E = \begin{pmatrix} 1 & 0 \\ 0 & 1 \end{pmatrix}$, we have*

$$V' = 1, \qquad \text{with } A = \begin{pmatrix} 0 & 1 \\ 0 & 0 \end{pmatrix}$$

corresponding to

$$V' = (1 - cV)^2, \qquad \text{with } A = \begin{pmatrix} -c & 1 \\ -c^2 & c \end{pmatrix}$$

The canonical variable V for the exponential system is thus a FLT of V for the Gaussian system, which has $V(z) = z$.

Proof: Following Prop. 5.2.4, we calculate:

$$EAE^{-1} = \begin{pmatrix} -c & 1 \\ -c^2 & c \end{pmatrix}$$

Then, using the correspondence, eq. (5.2.1), between the coefficient matrix and the Riccati equation, the result follows. ∎

5.3.2 Poisson-Bernoulli orbit

Here we start with the Poisson system as the simplest system with non-zero δ. We have the following, with proof similar to that for the Gaussian-exponential orbit. Here we are using the term 'Bernoulli system' to denote the various binomial systems.

5.3.2.1 Proposition. *Under the action of the group of FLT's the Poisson system maps to the Bernoulli system. Under the transformation* $E = \begin{pmatrix} 1 & 0 \\ c & 1 \end{pmatrix}$*, we have*

$$V' = 1 + 2\alpha V, \qquad \text{with } A = \begin{pmatrix} \alpha & 1 \\ 0 & -\alpha \end{pmatrix}$$

corresponding to

$$V' = 1 + 2(\alpha - c)V + c(c - 2\alpha)V^2, \qquad \text{with } A = \begin{pmatrix} \alpha - c & 1 \\ 2\alpha c - c^2 & c - \alpha \end{pmatrix}$$

The canonical variable V for the Bernoulli system is thus a FLT of V for the Poisson system, which has $V(z) = (e^{2\alpha z} - 1)/2\alpha$.

Since in general we have two parameters α, β, the two parameters α, c should be enough to determine the general system. This is so only if one permits an imaginary jump size 2α in the Poisson process, as we will now see. Call the desired parameters A, B. Then, from the above Proposition, we choose α and c to satisfy:

$$\alpha - c = A$$
$$c^2 - 2\alpha c = B$$

Solving these directly, or using the invariance of δ^2, we have

$$\alpha = \pm\sqrt{A^2 - B}, \qquad c = \pm\sqrt{A^2 - B} - A$$

which entails imaginary α for $A^2 < B$. In any case, the canonical variable V for all three of the binomial systems is a fractional linear transformation of the exponential function, allowing for imaginary values of α.

VI. Reproducing kernels

Now we look at reproducing kernels for the Bernoulli systems (Ch. 3, §2.2.2). For Bernoulli systems, we have the bases $\{J_n\}$ or $\{\phi_n\}$. Except for the binomial system, which is finite-dimensional, we have only weak reproducing kernels, i.e., the series do not converge to give analytic functions.

6.1 Definition. The kernel $K(x, y)$ is given by

$$K(x, y) = \sum_{n=0}^{\infty} \frac{J_n(x)J_n(y)}{\gamma_n}$$

We can see how this relates to the theory of this Chapter.

6.2 Proposition.   *The kernel $K(x,y)$ has the representation*

$$K(x,y) = \omega(R,y)1$$

where R is the raising operator for the corresponding Bernoulli system.

Proof: Starting with $\omega(y,x) = \sum (y^n/n!)\,\phi_n(x)$, we have

$$\omega(R,y)1 = \sum_{n=0}^{\infty} \frac{\phi_n(y)R^n 1}{n!}$$

$$= \sum_{n=0}^{\infty} \frac{\phi_n(y)J_n(x)}{n!}$$

$$= \sum_{n=0}^{\infty} \frac{J_n(x)J_n(y)}{\gamma_n}$$

via the relation $\phi_n = (n!/\gamma_n)J_n$.   ∎

Remark. It is interesting to combine this formulation with Prop. 2.1.1. Write

$$\int_{-\infty}^{\infty} e^{ay}\,\omega(b,y)\,p(dy) = e^{bV(a)+H(a)}$$

Following the above Proposition, we have

$$\int_{-\infty}^{\infty} e^{ay}\,\omega(R,y)1\,p(dy) = e^{H(a)}\,e^{V(a)R}\,1$$

On the left side, since we have the reproducing kernel, we get back the exponential, $\exp(ax)$. Expanding out the right side, we thus have

$$e^{ax} = e^{H(a)} \sum_{n=0}^{\infty} \frac{V(a)^n}{n!}\,J_n(x)$$

which is the basic expansion used in §I.

Now we put in a convergence factor, which will give us approximate reproducing kernels in analytic form.

6.3 Definition.   The kernel $K_\lambda(x,y)$ is given by

$$K_\lambda(x,y) = \sum_{n=0}^{\infty} \frac{\lambda^n J_n(x)J_n(y)}{\gamma_n}$$

In the remainder of this section, §VI, we will find the kernels K_λ for the various Bernoulli systems. In Chapter 7, we will see that the resulting formulas are closely related to (in fact, are essentially the same as) *addition formulas* for the matrix elements of the HW and sl(2) groups. Here we use group-theoretical methods in a direct approach. We start with the Gaussian system, as it is the simplest. In agreement with the formulation of Chapter 5, we bring back t as an explicit parameter.

6.1 GAUSSIAN SYSTEM

Here we use the fact that the polynomials are given as time-reversed moment polynomials corresponding to the generator $H(z) = z^2/2$. I.e., with $\gamma_n = n! \, t^n$,

6.1.1 Lemma. *For the Gaussian system we have the relation*

$$K_\lambda(x,y) = e^{-tH} \, e^{(\lambda xy/t) - (\lambda^2 x^2/2t)}$$

with $H(D) = D^2/2$.

Proof: From $H_n(x,t) = \exp(-tH)x^n$ follows

$$K_\lambda(x,y) = e^{-tH} \sum_{n=0}^{\infty} \frac{\lambda^n x^n H_n(y,t)}{n! \, t^n}$$

$$= e^{-tH} \, G_t(\lambda x/t, y)$$

where G_t is the generating function for the Hermite polynomials (Ch 5., Theorem 7.2.4). The result follows. ∎

Remark. In the proof we see from the form of the summation that this is the kernel ω for the Gaussian system. Thus,

$$\omega(x,y) = e^{(xy/t) - (x^2/2t)}$$

And we have the relation

$$K_\lambda(x,y) = e^{-tH} \, \omega(\lambda x, y)$$

Notice that we require the action of the exponential of $D^2/2$ on a function of $x^2/2$ (besides the term involving just x). This is the $N = 1$ realization of sl(2) with Δ as half the Laplacian, cf. Ch. 1, proof of Prop. 3.1.3 and eq. (3.1.2).

6.1.2 Theorem. *The kernel K_λ for the Gaussian system is*

$$(1 - \lambda^2)^{-1/2} \exp\left(\frac{\lambda^2 x^2 + \lambda^2 y^2 - 2\lambda xy}{2t(\lambda^2 - 1)} \right)$$

Proof: We use the realization $\Delta = D^2/2$, $\rho = xD + \frac{1}{2}$, $R = x^2/2$ of the sl(2) algebra. From the exponential commutation rule, Ch. 1, Prop. 3.3.2, we have

$$e^{-tD^2/2} \, e^{(\lambda xy/t) - (\lambda^2 x^2/2t)} =$$

$$\exp\left(\frac{-\lambda^2/t}{1 - \lambda^2} \frac{x^2}{2} \right) (1 - \lambda^2)^{-\rho} \exp\left(\frac{-t}{1 - \lambda^2} \frac{\lambda^2 y^2}{2t^2} \right) e^{(\lambda xy/t)}$$

Now we calculate, using the action $a^{xD} f(x) = f(ax)$, Ch. 1, Prop. 2.4.2 and the remark there:

$$(1 - \lambda^2)^{-\rho} e^{(\lambda xy/t)} = (1 - \lambda^2)^{-1/2} \exp\left(\frac{\lambda xy}{t(1 - \lambda^2)} \right)$$

And the result follows via the Lemma. ∎

This kernel is known as the *Mehler-Fock kernel* .

6.2 EXPONENTIAL SYSTEM

For the exponential system, we take $\alpha = 1$, so that with $\delta^2 = \alpha^2 - \beta$, $\beta = 1$ as well. Thus, $\gamma_n = n!\,(t)_n$. Now, with $y = 1$ in Prop. 3.2.1.4 of Chapter 4, the Laguerre polynomials are given by (see also Ch. 5, Prop. 6.4.2)

$$L_n(x,t) = e^{-\Delta}\,x^n = (-1)^n (t)_n\,{}_1F_1 \left(\begin{array}{c} -n \\ t \end{array} \middle| x \right) \tag{6.2.1}$$

Here we are using the realization of sl(2) as

$$\Delta = xD^2 + tD, \qquad \rho = t + 2xD, \qquad R = x \tag{6.2.2}$$

And we see immediately, then,

6.2.1 Lemma. *For the exponential system we have the relation*

$$K_\lambda(x,y) = e^{-\Delta} \sum_{n=0}^{\infty} \frac{\lambda^n x^n L_n(y,t)}{n!\,(t)_n}$$

Recall the functions $\mathcal{I}_t(x)$, Ch. 3, Prop. 2.4.1.2, here with t replacing c,

$$\mathcal{I}_t(x) = \sum_{n=0}^{\infty} \frac{x^n}{n!\,(t)_n} = {}_0F_1 \left(\begin{array}{c} - \\ t \end{array} \middle| x \right)$$

The functions $\mathcal{I}_t(xy)$ are eigenfunctions of Δ, satisfying $\Delta \mathcal{I}_t(xy) = y\mathcal{I}_t(xy)$. We have

6.2.2 Lemma. *The generating function relation*

$$e^{-\lambda x}\,\mathcal{I}_t(\lambda xy) = \sum_{n=0}^{\infty} \frac{\lambda^n x^n L_n(y,t)}{n!\,(t)_n}$$

Proof: Substitute the ${}_1F_1$ expression, eq. (6.2.1), for L_n in the series. Expanding out the ${}_1F_1$ and resumming gives the result (cf. Lemma 3.1.3). ∎

Remark. We see that the series is the ω-kernel for the exponential system. I.e.,

$$\omega(x,y) = e^{-x}\,\mathcal{I}_t(xy)$$

and the relation

$$K_\lambda(x,y) = e^{-\Delta}\,\omega(\lambda x, y)$$

follows.

We now compute K_λ.

6.2.3 Theorem. *For the exponential system, the kernel K_λ is given by*

$$(1 - \lambda)^{-t} \exp\left(-(x + y)\frac{\lambda}{1 - \lambda}\right) \mathcal{I}_t\left(\frac{\lambda x y}{(1 - \lambda)^2}\right)$$

Proof: As in the proof for the Gaussian system, we use the exponential commutation rule for sl(2), Ch. 1, Prop. 3.3.2, here in the realization given in eq. (6.2.2). Thus, via the above Lemmas, we have for K_λ:

$$e^{-\Delta} e^{-\lambda x} \mathcal{I}_t(\lambda x y) = \exp\left(\frac{-\lambda}{1 - \lambda} x\right) (1 - \lambda)^{-\rho} \exp\left(\frac{-\lambda}{1 - \lambda} y\right) \mathcal{I}_t(\lambda x y)$$

where we use the fact remarked above that the $\mathcal{I}_t$ are eigenfunctions of Δ. It remains to calculate the action of $(1 - \lambda)^{-\rho}$. With $\rho x^n = (t + 2n)x^n$, cf. eq. (6.2.2), we find

$$(1 - \lambda)^{-\rho} \mathcal{I}_t(\lambda x y) = \sum_{n=0}^{\infty} \frac{(\lambda y)^n (1 - \lambda)^{-t-2n} x^n}{n! \, (t)_n}$$

$$= (1 - \lambda)^{-t} \mathcal{I}_t\left(\frac{\lambda x y}{(1 - \lambda)^2}\right)$$

And the result follows. ∎

This kernel is known as the *Hardy-Hille kernel* .

6.3 POISSON SYSTEM

For the Poisson system, we have the Poisson-Charlier polynomials (cf. remark following Prop. 5.4.2 of Chapter 5 and Corollary 5.4.4 there). With $\gamma_n = n! \, t^n$, the kernel K_λ has the form

$$K_\lambda(x, y) = \sum_{n=0}^{\infty} \frac{\lambda^n P_n(x) P_n(y)}{n! \, t^n}$$

For this system, we just give the result, as it is essentially the same as the addition formula for the HW group, which we will study in Ch. 7, §4.1.

6.3.1 Theorem. *For the Poisson system, the kernel K_λ is given by*

$$(1 - \lambda)^{x+y} e^{\lambda t} \, {}_2F_0\left(\begin{array}{c} -x, -y \\ \underline{\quad} \end{array} \middle| \frac{\lambda}{t(1 - \lambda)^2}\right)$$

We call this the *Poisson-Charlier kernel* .

6.4 BERNOULLI SYSTEM

For the binomial systems, we will find the kernel for the general Bernoulli system. Ths idea is to use the gamma transform to build $_2F_1$ functions from $_1F_1$ functions. Since we know the kernel for the Laguerre polynomials, we can find the kernel for the general Bernoulli polynomials.

First we have a general fact.

6.4.1 Lemma. Let $\mathbf{a} = (a_1, \ldots, a_m)$ and $\mathbf{b} = (b_1, \ldots, b_n)$ denote numerator and denominator parameters for a $_mF_n$ hypergeometric function:

$$F_\lambda(x) = {}_mF_n\left({\mathbf{a} \atop \mathbf{b}} \,\middle|\, \lambda x\right)$$

With $e_s(x) = e^{sx}$, as in Lemma 4.1.2, we have the gamma transform

$$\Gamma_t[e_s \cdot F_\lambda](v) = (1 - sv)^{-t} \, {}_{m+1}F_n\left({\mathbf{a}, t \atop \mathbf{b}} \,\middle|\, \frac{\lambda v}{1 - sv}\right)$$

Proof: From Lemma 4.1.2, we have the transform

$$\Gamma_t[e_s \cdot f_k](v) = (1 - sv)^{-t}(t)_k \left(\frac{v}{1 - sv}\right)^k$$

where $f_k(x) = x^k$. The result follows via expanding the hypergeometric functions as series. ∎

We will use this Lemma to go from $_1F_1$ to $_2F_1$ functions. So now we derive a kernel for $_2F_1$ functions.

6.4.2 Lemma. We have the following $_2F_1$ kernel:

$$\sum_{n=0}^{\infty} \frac{w^n}{n!}(t)_n \, {}_2F_1\left({-n, X \atop t} \,\middle|\, v\right) \, {}_2F_1\left({-n, Y \atop t} \,\middle|\, v\right)$$

$$= (1 - w)^{X+Y-t}(1 - (1 - v)w)^{-X-Y} \, {}_2F_1\left({X, Y \atop t} \,\middle|\, \frac{v^2 w}{(1 - (1 - v)w)^2}\right)$$

Proof: Write the kernel for the exponential system, Theorem 6.2.3, in terms of $_1F_1$ hypergeometric functions:

$$\sum_{n=0}^{\infty} \frac{w^n(t)_n}{n!} \, {}_1F_1\left({-n \atop t} \,\middle|\, x\right) \, {}_1F_1\left({-n \atop t} \,\middle|\, y\right)$$

$$= (1 - w)^{-t} \exp\left(-(x + y)\frac{w}{1 - w}\right) \mathcal{I}_t\left(\frac{wxy}{(1 - w)^2}\right)$$

Now, set $A = w/(1 - w)$, $B = w/(1 - w)^2$. Recalling that $\mathcal{I}_t$ is a $_0F_1$ function, we take gamma transforms $\Gamma_X \Gamma_Y$ in the variables x and y respectively, where for clarity we keep the arguments explicit. Using Lemma 6.4.1, we have

$$\Gamma_X \Gamma_Y [e^{-A(x+y)} \, \mathcal{I}_t(Bxy)](v) = \Gamma_Y \left[e^{-Ay} (1 + Av)^{-X} \, {}_1F_1 \left(\begin{matrix} X \\ t \end{matrix} \,\middle|\, \frac{Bv}{1 + Av} y \right) \right]$$

$$= (1 + Av)^{-X-Y} \, {}_2F_1 \left(\begin{matrix} X, Y \\ t \end{matrix} \,\middle|\, \frac{Bv^2}{(1 + Av)^2} \right)$$

Substituting for A and B, the result follows. ∎

And for the Bernoulli systems we have

6.4.3 Theorem.   *For the general Bernoulli system the kernel K_λ is given by*

$$(1 - \lambda)^{-X-Y} (1 - \lambda(\alpha + \delta)/(\alpha - \delta))^{X+Y-t} \, {}_2F_1 \left(\begin{matrix} X, Y \\ t \end{matrix} \,\middle|\, \frac{4\delta^2 \lambda}{\beta(1 - \lambda)^2} \right)$$

where $X = (x + \delta t)/2\delta$, $Y = (y + \delta t)/2\delta$.

Proof: We have the form of the polynomials $J_n(x, t)$ from Ch. 5, Theorem 1.5.2.1:

$$J_n(x, t) = (-1)^n (t)_n (\alpha + \delta)^n \, {}_2F_1 \left(\begin{matrix} -n, X \\ t \end{matrix} \,\middle|\, \frac{2\delta}{\alpha + \delta} \right)$$

where $X = (x + \delta t)/2\delta$. With $\gamma_n = n! \, \beta^n (t)_n$, we have

$$K_\lambda(x, y) = \sum_{n=0}^{\infty} \frac{\lambda^n J_n(x, t) J_n(y, t)}{n! \, \beta^n (t)_n}$$

Substituting in for the J_n, we find a series as in Lemma 6.4.2, with

$$w = \frac{\alpha + \delta}{\alpha - \delta} \lambda, \qquad v = \frac{2\delta}{\alpha + \delta}$$

recalling the relation $\beta = (\alpha + \delta)(\alpha - \delta)$. The result follows, then, from Lemma 6.4.2. ∎

This kernel is the *Meixner kernel* .

This concludes our study of the Bernoulli systems.

VII. Exercises

7.1 EXERCISES

1. Fill in details of Proposition 1.1.3.

2. Show that the moment operators $\eta_k(z,t)$ are the moments for the exponential family μ_z.

3. Write the Rodrigues formula corresponding to each of the Bernoulli systems of Chapter 5.

4. Check details of the statements of §1.4.

5. From equation (2.1) find the ω-kernel using the generating functions for the Poisson and Gaussian systems.

6. Fill in the details of Theorem 2.2.

7. Deduce Theorem 2.1.2 from Theorem 1.3.1.

8. In Lemma 3.1.1 verify directly the solution claimed.

9. Show Lemma 3.1.3. Deduce Theorem 3.1.4.

10. Calculate $m_n(y)$ from Definition 3.2.1, for $n = 0, 1, 2, 3$.

11. Using Proposition 3.2.3 find a formula for $m_n(y)$ for the Gaussian.

12. Show details of Lemma 4.1.2.

13. Write out the proof of Theorem 4.3.1 for the Gaussian system, $H(z) = z^2/2$.

14. Verify the correspondence between FLT's and matrices under composition.

15. Verify Proposition 5.1.2.

16. Show details for Proposition 5.2.1.

17. Use Proposition 5.2.2 to find the solution to the Riccati equation $V' = 1 - cV^2$ corresponding to the matrix $A = \begin{pmatrix} 0 & 1 \\ c & 0 \end{pmatrix}$ with the initial conditions $(Y_0, y_0) = (0, 1)$.

18. Referring to the remark following Proposition 5.2.4, find the set of matrices $\begin{pmatrix} \alpha & \beta \\ \gamma & \delta \end{pmatrix}$ commuting with the matrix A of Problem 17. Discuss.

19. Verify the transformations claimed in Propositions 5.3.1.1 and 5.3.2.1.

20. In §§6.1, 6.2 verify details of the calculations involving exponential commutation rules.

21. Try to prove Theorem 6.3.1.

22. Check the details in Lemma 6.4.2 and Theorem 6.4.3.

Chapter 7   MATRIX ELEMENTS

First we look at the matrix elements for the HW, oscillator, and SL(2) groups corresponding to coordinates of the second kind, i.e., for group elements as in Ch.1, §§2.3, 3.4. Then, using the splitting formulas from Ch.1, §IV, we find matrix elements for the group elements in the form e^{sX}, where X is one of the X operators from the Bernoulli systems. These generalize the moment generating functions, which are the vacuum-to-vacuum matrix elements, $\langle e^{sX} \rangle$. Then, via the Riccati equation, we find addition formulas for V. This leads to the function Ψ, which arises as well in the *coherent state representation*, §III, of the operator X. Via the coherent state representation, we will see the rôle of the kernel ω as a generalized exponential. In the concluding section, we find addition formulas for the functions arising in the general matrix elements. For the one-parameter subgroups of the form e^{sX} these yield a representation of the translation group, with the coordinates of the first kind as parameters. The resulting formulas correspond with the addition formulas for V, via the function Ψ.

Remark. In this chapter we return to the conventions of Chapter 5, with t as an explicit parameter in the Bernoulli systems, in contrast to Chapter 6 (see §III below). We will be generally working in a Fock space Φ, which may be of HW or sl(2) type. In general, we follow the conventions of Chapter 5 regarding the operators on the Fock space, i.e., we are always working in the context of Bernoulli systems, whether expressed or implied. Thus, e.g., the basis ψ_n is explicitly realized (in general) as the polynomial basis $\{ J_n(x,t) \}$ of the Bernoulli systems.

I.   Matrix elements

We have the orthogonal basis $\{ \psi_n \}$ with squared norms γ_n. The matrix elements of an operator E are defined as

$$M_{mn}(E) = \langle E\psi_n, \psi_m \rangle$$

and we have the formal expansion:

$$E\psi_n = \sum_{m=0}^{\infty} M_{mn}(E)\, \psi_m / \gamma_m \tag{1.1}$$

which in the cases we are considering will be explicitly given as analytic expressions. The matrix elements satisfy the composition rule, the *addition formula*

$$M_{mn}(E_1 E_2) = \sum_j M_{mj}(E_1)\gamma_j^{-1} M_{jn}(E_2) \tag{1.2}$$

as a consequence of the expansion (1.1).

1.1 HW GROUP MATRIX ELEMENTS

We have the basis $\psi_n = R^n \psi_0$ with squared norms $\gamma_n = n! \, h^n$. Consider the matrix elements for the HW group in the form

$$\langle e^{aR} e^{bL} \psi_n, \psi_m \rangle$$

with $L = hV$. Note that $\exp(ch)$ would come right out as a common factor.

1.1.1 Proposition. *For the HW group, the matrix elements have the form:*

$$\langle e^{aR} e^{bL} \psi_n, \psi_m \rangle = (ah)^m (bh)^n \, {}_2F_0 \left(\begin{array}{c} -m, -n \\ \rule{1cm}{0.4pt} \end{array} \middle| \frac{1}{abh} \right)$$

Proof: Using the fact that R and L are adjoint, via the matrix elements given in Ch.1, Prop. 2.2.3.1:

$$\langle e^{aR} e^{bL} \psi_n, \psi_m \rangle = \langle e^{bL} \psi_n, e^{aL} \psi_m \rangle$$

$$= \langle \sum_{k=0}^{n} \binom{n}{k} b^{n-k} h^{n-k} \psi_k, \sum_{j} \binom{m}{j} a^{m-j} h^{m-j} \psi_j \rangle$$

which gives

$$\sum_{k=0}^{n} \binom{n}{k} \binom{m}{k} b^{n-k} a^{m-k} h^{n+m-2k} h^k k! = (ah)^m (bh)^n \, {}_2F_0 \left(\begin{array}{c} -m, -n \\ \rule{1cm}{0.4pt} \end{array} \middle| \frac{1}{abh} \right)$$

$\blacksquare$

Now we consider the X-operator for the Gaussian system, Ch. 5, §7.2, corresponding to the HW algebra.

1.1.2 Proposition. *Let $X = R + tV$ be an element of the HW algebra. The matrix elements for the group generated by X are given by:*

$$\langle e^{sX} \psi_n, \psi_m \rangle = e^{s^2 t/2} \, (st)^{m+n} \, {}_2F_0 \left(\begin{array}{c} -n, -m \\ \rule{1cm}{0.4pt} \end{array} \middle| \frac{1}{s^2 t} \right)$$

Proof: By the splitting formula, Ch. 1, Prop. 4.1.1,

$$e^{sX} = \exp(s(R + tV)) = e^{sR} \, e^{stV} \, e^{s^2 t/2}$$

Now apply Prop. 1.1.1, with $a = b = s$ and $h = t$ to get the result. $\blacksquare$

1.2 OSCILLATOR GROUP MATRIX ELEMENTS

For the oscillator group, we use the basis elements R, RV, L with $L = hV$. We have, as for the HW case,

1.2.1 Proposition. *The matrix elements for the oscillator group are*

$$\langle e^{aR} e^{cRV} e^{bL} \psi_n, \psi_m \rangle = (ah)^m (bh)^n \, {}_2F_0 \left(\begin{matrix} -m, -n \\ {}-\!\!-\!\!- \end{matrix} \, \middle| \, \frac{e^c}{abh} \right)$$

Proof: As in the proof of Prop. 1.1.1,

$$\langle e^{aR} e^{cRV} e^{bL} \psi_n, \psi_m \rangle = \langle e^{cRV} e^{bL} \psi_n, e^{aL} \psi_m \rangle$$

$$= \sum_{k=0}^{n} \binom{n}{k} \binom{m}{k} b^{n-k} a^{m-k} h^{n+m-2k} h^k k! \, e^{ck}$$

where the factor $\exp(ck)$ comes via the number operator RV. The result follows. ∎

The X-operator for the Poisson system, Ch. 5, §5.2, corresponds to the oscillator algebra. And we have

1.2.2 Proposition. *Let $X = R + \alpha RV + tV$ be an element of the oscillator algebra. The matrix elements for the group generated by X are given by:*

$$\langle e^{sX} \psi_n, \psi_m \rangle = e^{tH(s)} (tV(s))^{m+n} \, {}_2F_0 \left(\begin{matrix} -n, -m \\ {}-\!\!-\!\!- \end{matrix} \, \middle| \, \frac{e^{\alpha s}}{tV(s)^2} \right)$$

where $V(s) = (e^{\alpha s} - 1)/\alpha$ and $H(s) = (e^{\alpha s} - 1 - \alpha s)/\alpha^2$.

Proof: By the splitting formula, Ch. 1, Prop. 4.2.1,

$$e^{sX} = \exp(s(R + \alpha RV + tV)) = e^{tH(s)R} e^{V(s)R} e^{\alpha s RV} e^{V(s)L}$$

where $L = tV$ here. Now apply Prop. 1.2.1, with $a = b = V(s)$, $c = \alpha s$, and $h = t$. ∎

Remark. Note that the argument in the ${}_2F_0$ may be expressed as

$$\frac{e^{\alpha s}}{tV(s)^2} = \frac{(\alpha/2)^2}{t \sinh^2(\alpha s/2)}.$$

1.3 SL(2) GROUP MATRIX ELEMENTS

For the SL(2) matrix elements, we use the standard basis, in terms of bosons R, V: $R = R$, $L = \beta\Delta = \beta(tV + RV^2)$, $\rho = [\Delta, R] = t + 2RV$, here with t instead of c, as usual for Bernoulli systems.

1.3.1 Proposition. *The matrix elements for SL(2) are:*

$$\langle e^{aR} c^\rho e^{bL} \psi_n, \psi_m \rangle = c^t(t)_n(t)_m (a\beta)^m (b\beta)^n \, {}_2F_1 \left(\begin{matrix} -m, -n \\ t \end{matrix} \, \middle| \, \frac{c^2}{ab\beta} \right)$$

Proof: Using the matrix elements from Ch. 1, Prop. 3.3.2.1,

$$\langle e^{aR} c^\rho e^{bL} \psi_n, \psi_m \rangle =$$

$$\sum_{k,j} \langle c^\rho \binom{n}{k} \binom{n+t-1}{n-k} (n-k)!(\beta b)^{n-k} \psi_k, \binom{m}{j} \binom{m+t-1}{m-j} (m-j)!(\beta a)^{m-j} \psi_j \rangle$$

$$= \sum_{k=0}^{n} \binom{n}{k} \frac{\Gamma(n+t)}{\Gamma(k+t)} \binom{m}{k} \frac{\Gamma(m+t)}{\Gamma(k+t)} c^{t+2k} \beta^{n+m-k} b^{n-k} a^{m-k} k! \, (t)_k$$

which yields the result after pulling out the indicated factors. ∎

Now for the X-operator for the general Bernoulli system, Ch. 5, §1.2.

1.3.2 Proposition. *Let $X = R + \alpha\rho + \beta\Delta$ be an element of the sl(2) algebra. The matrix elements for the group generated by X are:*

$$\langle e^{sX} \psi_n, \psi_m \rangle = e^{tH(s)} \frac{\gamma_m \gamma_n}{m! \, n!} V(s)^{m+n} \, {}_2F_1 \left(\begin{matrix} -m, -n \\ t \end{matrix} \, \middle| \, \frac{\pi(V(s))}{\beta V(s)^2} \right)$$

with $V(s) = \tanh \delta s / (\delta - \alpha \tanh \delta s)$, $H(s) = \log(\delta \operatorname{sech} \delta s / (\delta - \alpha \tanh \delta s))$, and the squared norms $\gamma_n = n! \, \beta^n(t)_n$. The characteristic polynomial $\pi(v) = 1 + 2\alpha v + \beta v^2$.

Proof: As in the proof of Ch. 5, Theorem 1.2.1, we have the splitting formula

$$e^{sX} = e^{V(s)R} e^{\rho H(s)} e^{V(s)L}$$

with V and H as stated. In Prop. 1.3.1 substitute $a = b = V(s)$ and $c = \exp(H(s))$. Recall in the proof of the splitting formula, Ch. 1, eq. (4.3.2) (also see the proof of Ch. 5, Theorem 1.5.1.1), $V' = \exp(2H)$. Combining with the Riccati equation $V' = \pi(V)$, the result follows. ∎

Remark. Note that here the argument of the ${}_2F_1$ may be expressed as

$$\frac{\pi(V(s))}{\beta V(s)^2} = \frac{\delta^2 / \beta}{\sinh^2 \delta s}$$

II.   Addition formulas and Riccati equations

Here we will see that solutions of the Riccati equations satisfy addition formulas. This leads to a particular function Ψ characteristic of the Bernoulli systems that will appear prominently in the coherent state representations of §III.

2.1   ADDITION FORMULAS

Recall, Ch. 6, §V, that we can generate solutions to the Riccati equations via fractional linear transformations (FLT's). The simplest Riccati equation, $V' = 1$, has the solution $V(z) = z$, which satisfies the addition formula

$$V(a + b) = V(a) + V(b)$$

If we apply FLT's to both equations, the Riccati equation and the addition formula, we will find an addition formula for the solution to the more general Riccati equation. As we saw in Ch. 6, §5.3, starting from the Poisson system we can generate the general Riccati equation for the Bernoulli systems.

2.1.1 Lemma.   *The solution to $V' = 1 + 2\alpha V$, $V(0) = 0$, satisfies the addition formula*

$$V(a + b) = V(a) + V(b) + 2\alpha V(a)V(b) \qquad (2.1.1)$$

Proof: This follows directly from $V(z) = (e^{2\alpha z} - 1)/2\alpha$.   ∎

Now we find for the general case

2.1.2 Theorem.   *The function $V(z)$ satisfies the Riccati equation*

$$V' = 1 + 2\alpha V + \beta V^2, \qquad V(0) = 0$$

if and only if it satisfies the addition formula

$$V(a + b) = \frac{V(a) + V(b) + 2\alpha V(a)V(b)}{1 - \beta V(a)V(b)}$$

with the conditions $V(0) = 0$, $V'(0) = 1$.

Proof: From the addition formula to the differential equation is straightforward, using $V(0) = 0$, $V'(0) = 1$:

$$\frac{V(a + b) - V(a)}{b} = \frac{V(b)}{b} \frac{1 + 2\alpha V(a) + \beta V(a)^2}{1 - \beta V(a)V(b)}$$
$$\rightarrow V'(0)\,(1 + 2\alpha V(a) + \beta V(a)^2), \qquad \text{as } b \rightarrow 0$$

To go from the Riccati equation to the addition formula, we use the correspondence via FLT's. The Lemma is the case $\beta = 0$. Now, as in Ch.6, Prop. 5.2.4, we write, according to the action of the transformation $\left(\begin{smallmatrix} 1 & 0 \\ c & 1 \end{smallmatrix}\right)$:

$$v = \frac{V}{cV + 1}, \qquad V = \frac{v}{1 - cv}$$

As seen there, v satisfies

$$v' = 1 + 2(\alpha - c)v + c(c - 2\alpha)v^2$$

To check the addition formula, substitute in for $V(a)$, $V(b)$, $V(a+b)$ in eq. (2.1.1). After some rearranging of terms, one arrives at the stated form, with parameters $\alpha - c$, $c(c - 2\alpha)$, in accordance with the Riccati equation for v. Replacing these by α, β, the result follows. ∎

2.2 FUNCTION Ψ

Now we can define the function Ψ which is intimately related to the structure of Bernoulli systems, as we will see in the next section, §III.

2.2.1 Definition. Given the parameters α, β, we denote

$$\Psi(a, b) = \frac{a + b + 2\alpha ab}{1 - \beta ab}$$

Thus, we write the above Theorem:

$$V(a + b) = \Psi(V(a), V(b))$$

And we have

2.2.2 Corollary. *With U the inverse function to V, we have the addition formula*

$$V(U(a) + U(b)) = \Psi(a, b)$$

Compare with the *cocycle identity*, Theorem 4.2.5 of Chapter 6, which can be written in either of the forms

$$H(a + b) - H(a) - H(b) = \phi(V(a)V(b))$$

or

$$H(U(a) + U(b)) - H(U(a)) - H(U(b)) = \phi(ab)$$

where throughout this Chapter, we have

$$\phi(x) = -\log(1 - \beta x)$$

with the t scaled out, since we are working explicitly with the scaled generator tH, as in Chapter 5.

III.   Coherent states and coherent state representations

The term *coherent states* derives from physics, referring to quantum states that evolve much like classical states; they have minimal spreading (dispersion). Here we use the algebraic approach concordant with that of Perelomov [38]. In this approach, the coherent states are group elements applied to an appropriate cyclic vector (vacuum state). Another way to think of these states, as we will see below, is as generating functions.

3.1   COHERENT STATES

As we have seen, Ch. 5, §1.4, we have for the group elements generated by X:

$$e^{zX}\, \Omega = e^{tH(z)}\, e^{V(z)R}\, \Omega$$

Up to the scalar factor $\exp(tH(z))$ and the change of variables $z \to U(v)$, this is the generating function

$$e^{vR}\, 1 = e^{xU(v)-tM(v)}$$

for the Bernoulli systems (with Ω taken to be the constant function 1). So

3.1.1 Definition.   The *coherent state* ψ_a is the generating function

$$\psi_a = e^{aR}\, \Omega$$

This agrees with the intuition that all modes (states ψ_n) should be present in a coherent state.

Remark. Typically we use ψ_a, ψ_b to denote these states.

Denoting the inner product between the states ψ_a, ψ_b by γ_{ab}, we have

3.1.2 Proposition.   *The coherent states satisfy*

$$\gamma_{ab} = \langle \psi_a, \psi_b \rangle = e^{t\phi(ab)}$$

with $\phi(x) = -\log(1 - \beta x)$.

Proof: This is the same calculation as in Theorem 4.3.1 of Chapter 6. See Ch. 6, eq. (4.3.2), with $V(a) \to a$, $V(b) \to b$. The result follows.   ∎

Remark. Note that we require $|\beta ab| < 1$. For ψ_a to be in the Fock space Φ, we have the condition $|\beta a^2| < 1$. Then we can see directly that if $\psi_a, \psi_b \in \Phi$, their inner product is well-defined:

$$|\beta ab| = \sqrt{(\beta a^2)(\beta b^2)} < 1$$

For operators on Φ we have

3.1.3 Definition. The *coherent state representation* or *CSR* of an operator Q is

$$\langle Q \rangle_{ab} = \frac{\langle Q\psi_a, \psi_b \rangle}{\langle \psi_a, \psi_b \rangle}$$

Notice that for $a \neq b$, we do not claim to have a true expectation, since positivity will be lacking in general. We do, however, have linearity and the relation $\langle I \rangle_{ab} = 1$ for the identity operator.

The matrix elements of operators with respect to coherent states act much like Fourier transforms. For us an important feature is:

3.1.4 Proposition. *The unnormalized CSR, $\gamma_{ab}\langle Q \rangle_{ab}$, is the generating function for the matrix elements with respect to the basis states ψ_n.*

Proof: Expanding the states ψ_a, ψ_b, we have

$$\gamma_{ab}\langle Q \rangle_{ab} = \langle Q\psi_a, \psi_b \rangle$$
$$= \sum_{n=0}^{\infty} \sum_{m=0}^{\infty} \frac{a^n b^m}{n!\, m!} \langle Q\psi_n, \psi_m \rangle$$

as required. ∎

Thus, the CSR is the generating function for the matrix elements we have been considering in §I.

3.2 CSR'S OF THE BASIC OPERATORS

To get an idea of the CSR as a transform, we will find the CSR's of the basic operators involved with the Bernoulli systems. We use the basic technique of differentiating with respect to parameters.

3.2.1 Lemma. *For the operators R, $N = \alpha\rho$, $L = \beta\Delta$, we have the CSR's*

$$\langle R \rangle_{ab} = \beta t\, \frac{b}{1 - \beta ab}\,, \qquad \langle L \rangle_{ab} = \beta t\, \frac{a}{1 - \beta ab}$$

and

$$\langle N \rangle_{ab} = \alpha t\, \frac{1 + \beta ab}{1 - \beta ab}$$

Proof: First, for R we have, via Prop. 3.1.2

$$\gamma_{ab} \langle R \rangle_{ab} = \frac{\partial}{\partial a} \langle e^{aR} \,\Omega, e^{bR} \,\Omega \rangle$$

$$= \frac{\partial}{\partial a} (1 - \beta ab)^{-t}$$

$$= \frac{\beta tb}{1 - \beta ab} \,\gamma_{ab}$$

And since R and L are adjoints,

$$\gamma_{ab} \langle L \rangle_{ab} = \frac{\beta ta}{1 - \beta ab} \,\gamma_{ab}$$

The action $\rho\psi_n = (t + 2n)\psi_n$ translates into

$$\rho\psi_a = \left(t + 2a \frac{\partial}{\partial a} \right) \psi_a$$

And for the CSR:

$$\gamma_{ab} \langle \rho \rangle_{ab} = \left(t + 2a \frac{\partial}{\partial a} \right) \gamma_{ab}$$

$$= \left(t + \frac{2ab\beta t}{1 - \beta ab} \right) \gamma_{ab}$$

$$= t \frac{1 + \beta ab}{1 - \beta ab} \,\gamma_{ab}$$

$\blacksquare$

Now we have for the X-operator

3.2.2 Theorem.   *The CSR of the operator X is*

$$\langle X \rangle_{ab} = \alpha t + \beta t \,\Psi(a, b)$$

with the function Ψ given in Definition 2.2.1.

Proof: From the Lemma we have

$$\langle X \rangle_{ab} = \langle R + N + L \rangle_{ab}$$

$$= t \frac{\beta(a + b + \alpha ab) + \alpha}{1 - \beta ab}$$

while

$$\alpha t + \beta t \,\Psi(a, b) = t \frac{\alpha(1 - \beta ab) + \beta(a + b + 2\alpha ab)}{1 - \beta ab}$$

$$= t \frac{\alpha + \beta(a + b + \alpha ab)}{1 - \beta ab}$$

in agreement. $\blacksquare$

Combining the theory of Chapter 6 with the results of this Chapter, we have now some principal results.

3.2.3 Theorem. *The CSR's of the basis polynomials in the operator X are given by:*

1. For the basis $\{\,\phi_n\,\}$:
$$\langle\phi_n(X)\rangle_{ab} = \Psi(a,b)^n$$

2. For the basis $\{\,J_n\,\}$:
$$\langle J_n(X,t)\rangle_{ab} = \beta^n(t)_n\,\Psi(a,b)^n$$

Proof: By Theorem 1.2.1 of Chapter 6, we have

$$\int_{-\infty}^{\infty} e^{zx}\,\phi_n(x)\,p_t(dx) = V(z)^n e^{tH(z)}$$

Thus,

$$
\begin{aligned}
\gamma_{ab}\langle\phi_n\rangle_{ab} &= e^{-t(M(a)+M(b))} \int_{-\infty}^{\infty} e^{x(U(a)+U(b))}\,\phi_n(x)\,p_t(dx) \\
&= [V(U(a)+U(b))]^n \, \exp\{t[H(U(a)+U(b)) - H(U(a)) - H(U(b))]\} \\
&= \Psi(a,b)^n \, e^{t\phi(ab)}
\end{aligned}
$$

by Corollary 2.2.2 and the cocycle identity, Ch. 6, Theorem 4.2.5. The result follows via Prop. 3.1.2. ∎

Now we have the *exponential formula* for ω.

3.2.4 Corollary. *The CSR of the ω-kernel:*

$$\langle\omega(y,X)\rangle_{ab} = e^{y\Psi(a,b)}$$

Proof: This follows from the Theorem and Ch. 6, Def. 2.1:

$$\omega(y,x) = \sum_{n=0}^{\infty}(y^n/n!)\,\phi_n(x)$$

■

For the group generated by X we have

3.2.5 Theorem. *The CSR's of the group elements e^{sX} :*

$$\langle e^{sX}\rangle_{ab} = e^{tH(s)} \, \exp\big[t\,\phi\big(V(s)\Psi(a,b)\big)\big]$$

Proof: As in the proof of Theorem 3.2.3, we have

$$\gamma_{ab}\langle e^{sX}\rangle_{ab} = \exp\Big\{t\big[H(s+U(a)+U(b)) - H(U(a)) - H(U(b))\big]\Big\}$$

$$= \exp\Big\{t\big[H(s) + H(U(a)+U(b)) - H(U(a)) - H(U(b)) + \phi(V(s)\Psi(a,b))\big]\Big\}$$

$$= \exp\Big\{t\big[H(s) + \phi(V(s)\Psi(a,b)) + \phi(ab)\big]\Big\}$$

by two applications of the cocycle identity and one application of Corollary 2.2.2.
∎

This is thus the generating function for the matrix elements $\langle e^{sX}\psi_n, \psi_m\rangle$ found in §I. And we see that this is the 'full version' of the moment generating function, which is the special case $a = b = 0$.

3.3 INTERPRETATIONS OF MATRIX ELEMENTS AND CSR'S

Here we will make explicit some of the analytic content of the matrix elements and CSR's.

1. The matrix element $\langle e^{sX}\psi_n, \psi_m\rangle$ is the generating function for the action of multiplication by x^l, say, on the basis $\{J_n(x,t)\}$:

$$x^l J_n(x,t) = \sum_m a_{ln}^m J_m(x,t) \tag{3.3.1}$$

Thus,

3.3.1 Proposition.   *The coefficients a_{ln}^m in the expansion*

$$x^l J_n(x,t) = \sum_m a_{ln}^m J_m(x,t)$$

are given by

$$a_{ln}^m = \gamma_m^{-1}\left(\frac{\partial}{\partial s}\right)^l\bigg|_{s=0} \langle e^{sX} J_n(X,t), J_m(X,t)\rangle$$

Proof: Multiplying by $s^l/l!$ and summing in eq. (3.3.1) gives

$$e^{sx} J_n(x,t) = \sum_{l,m} \frac{s^l}{l!} a_{ln}^m J_m(x,t)$$

The result follows by taking inner products with J_m.   ∎

2. The CSR $\gamma_{ab}\langle J_n\rangle_{ab}$ is the generating function for the *linearization formula* for the product $J_l J_m$:

$$J_l(x,t) J_m(x,t) = \sum_n c_{lm}^n J_n(x,t) \tag{3.3.2}$$

And we have

3.3.2 Proposition.   *The coefficients c_{lm}^n in the linearization formula*

$$J_l(x,t)J_m(x,t) = \sum_n c_{lm}^n J_n(x,t)$$

are given by

$$c_{lm}^n = \gamma_n^{-1} \frac{\partial^{l+m}}{\partial a^l \partial b^m} \bigg|_{a=b=0} \gamma_{ab} \langle J_n(X,t)\rangle_{ab}$$

Proof: Multiplying by $a^l b^m / l!\, m!$ in eq.(3.3.2) and summing yields

$$\psi_a \psi_b = \sum_{l,m,n} \frac{a^l b^m}{l!\, m!} c_{lm}^n J_n(x,t)$$

Taking inner products with J_n, the result follows.   ∎

Remark. We see as well that these coefficients give essentially the integral of the product of three polynomials:

$$\int_{-\infty}^{\infty} J_l(x,t)J_m(x,t)J_n(x,t)\, p_t(dx) = \gamma_n\, c_{lm}^n$$

from which permutation symmetry in the indices l, m, n is evident.

3. The CSR $\langle J_n \rangle_{ab}$ may be considered as giving the Fourier coefficients of the expansion of the product $\psi_a \psi_b$:

$$\frac{\psi_a \psi_b}{\langle \psi_a, \psi_b \rangle} = \sum_{n=0}^{\infty} \frac{\langle J_n \rangle_{ab}}{\gamma_n} J_n$$

We have from Theorem 3.2.3:

$$\frac{\psi_a \psi_b}{\langle \psi_a, \psi_b \rangle} = \sum_{n=0}^{\infty} \frac{\Psi(a,b)^n}{n!} J_n$$

The right side is the generating function $G_t(\Psi, x)$:

$$\sum_{n=0}^{\infty} \frac{\Psi(a,b)^n}{n!} J_n(x,t) = \exp\left[xU(\Psi(a,b)) - tM(\Psi(a,b))\right]$$

$$= \exp\left[x(U(a) + U(b)) - tH(U(a) + U(b))\right]$$

by Corollary 2.2.2. And this is again the cocycle identity with the identification $\langle \psi_a, \psi_b \rangle = \gamma_{ab} = e^{t\phi(ab)}$ via Prop. 3.1.2.

IV.   Addition formulas for matrix elements of the group

In conjunction with the group laws, Ch.1 §§2.3, 3.4, we can derive addition formulas for the matrix elements using eq. (1.2). For the matrix elements of $\exp(sX)$, $M_{mn}(s)$, say, we have the representation of the translation group

$$M_{mn}(s + s') = \sum_j M_{mj}(s)\gamma_j^{-1} M_{jn}(s')$$

which we will see is closely related to the addition formula for V, i.e., the function Ψ.

4.1   HEISENBERG GROUP

Since the terms involving $\exp(ch)$, e.g., will factor out, we use the group law in the form, cf. Ch.1, Prop. 2.3.2:

$$g(a, b, 0)\, g(A, B, 0) = g(a + A, b + B, Ab) \tag{4.1.1}$$

We can as well take $h = 1$, as this amounts to rescaling the b and B variables. Thus, we have the squared norms $\gamma_n = n!$.

4.1.1 Theorem.   *From the HW group, we have the addition formula:*

$$\sum_k \frac{w^k}{k!}\, {}_2F_0 \left(\begin{array}{c} -m, -k \\ \underline{\quad} \end{array} \middle| \frac{x}{w} \right) {}_2F_0 \left(\begin{array}{c} -k, -n \\ \underline{\quad} \end{array} \middle| \frac{y}{w} \right) =$$

$$e^w\, (1 + x)^m (1 + y)^n\, {}_2F_0 \left(\begin{array}{c} -m, -n \\ \underline{\quad} \end{array} \middle| w^{-1}\, \frac{x}{1 + x}\, \frac{y}{1 + y} \right)$$

Proof: From Prop. 1.1.1 the left side of eq. (4.1.1) yields, pulling through factors involving a, b, A, B:

$$\sum_k a^m (bA)^k B^n\, {}_2F_0 \left(\begin{array}{c} -m, -k \\ \underline{\quad} \end{array} \middle| \frac{1}{ab} \right) \frac{1}{k!}\, {}_2F_0 \left(\begin{array}{c} -k, -n \\ \underline{\quad} \end{array} \middle| \frac{1}{AB} \right)$$

Corresponding to the right side of eq. (4.1.1), we have

$$e^{Ab}\, (a + A)^m (b + B)^n\, {}_2F_0 \left(\begin{array}{c} -m, -n \\ \underline{\quad} \end{array} \middle| \frac{1}{(a + A)(b + B)} \right)$$

To get the result in the form stated, substitute:

$$w = Ab, \qquad x/w = 1/ab, \qquad y/w = 1/AB$$

∎

With this result as formulated, one sees by comparison with Prop. 1.2.1, that essentially the same general formula results from the matrix elements of the oscillator group.

4.2 SL(2) GROUP

Now we find the addition formula corresponding to the group SL(2). Taking $b = B = 1$, in the group law for SL(2), Ch.1, Prop. 3.4.1, hence replacing the variables c, C there by b, B, we have:

$$g(a, 1, b)\, g(A, 1, B) = g\left(a + \frac{A}{1 - Ab}, \frac{1}{1 - Ab}, B + \frac{b}{1 - Ab}\right) \qquad (4.2.1)$$

We can as well take $\beta = 1$, since as for the HW case, this amounts to rescaling the variables, as we saw in calculating the matrix elements, Prop. 1.3.1. We have

4.2.1 Theorem. *From the SL(2) matrix elements, we have the addition formula:*

$$\sum_k \frac{w^k}{k!}\, (t)_k\, {}_2F_1\left(\begin{matrix} -m, -k \\ t \end{matrix} \,\middle|\, x\right)\, {}_2F_1\left(\begin{matrix} -k, -n \\ t \end{matrix} \,\middle|\, y\right) =$$

$$(1 - w)^{-t-m-n}(1 - (1 - x)w)^m(1 - (1 - y)w)^n \times$$

$${}_2F_1\left(\begin{matrix} -m, -n \\ t \end{matrix} \,\middle|\, \frac{xyw}{(1 - (1 - x)w)(1 - (1 - y)w)}\right)$$

Proof: From Prop. 1.3.1, with $\beta = 1$, $c = C = 1$, we get

$$\sum_k (t)_k (t)_m (t)_k (t)_n a^m (bA)^k B^n\, {}_2F_1\left(\begin{matrix} -m, -k \\ t \end{matrix} \,\middle|\, \frac{1}{ab}\right) \frac{1}{(t)_k\, k!}\, {}_2F_1\left(\begin{matrix} -k, -n \\ t \end{matrix} \,\middle|\, \frac{1}{AB}\right)$$

corresponding to the left side of eq. (4.2.1). And for the right side, now with $c = 1/(1 - Ab)$ in Prop. 1.3.1:

$$(1 - Ab)^{-t}(t)_n (t)_m \left(a + \frac{A}{1 - Ab}\right)^m \left(B + \frac{b}{1 - Ab}\right)^n \times$$

$${}_2F_1\left(\begin{matrix} -m, -n \\ t \end{matrix} \,\middle|\, \frac{1}{(a + A - aAb)(B + b - ABb)}\right)$$

To get the result stated, make the substitutions:

$$w = Ab, \qquad x = 1/ab, \qquad y = 1/AB$$

after cancelling common factors. ∎

Compare with Ch. 6, Lemma 6.4.2.

4.3 REPRESENTATIONS OF THE TRANSLATION GROUP WITH GENERATOR X

Let us look at the addition formulas for the group elements of the form e^{sX} directly. Then we will compare with the addition formulas of §§4.1, 4.2. We consider the oscillator group, corresponding to the X-operator of the Poisson system (since $\alpha = 0$ recovers the HW case) and then SL(2).

4.3.1 Translations generated by X in the oscillator group

From Prop. 1.2.2, we have, with V and H for the Poisson system

$$M_{mn}(s) = e^{tH(s)} (tV(s))^{m+n} {}_2F_0\left(\begin{matrix} -n, -m \\ \underline{} \end{matrix} \,\middle|\, \frac{e^{\alpha s}}{tV(s)^2} \right)$$

We know that the matrix elements are a representation of the translation group:

$$M_{mn}(s + s') = \sum_j M_{mj}(s)(t^j j!)^{-1} M_{jn}(s')$$

Let us write this out explicitly in terms of the ${}_2F_0$ functions:

$$e^{tH(s+s')} (tV(s + s'))^{m+n} {}_2F_0\left(\begin{matrix} -n, -m \\ \underline{} \end{matrix} \,\middle|\, \frac{e^{\alpha(s+s')}}{tV(s + s')^2} \right) =$$

$$e^{t(H(s)+H(s'))} (tV(s))^m (tV(s'))^n \times$$

$$\sum_j \frac{(tV(s)V(s'))^j}{j!} \, {}_2F_0\left(\begin{matrix} -m, -j \\ \underline{} \end{matrix} \,\middle|\, \frac{e^{\alpha s}}{tV(s)^2} \right) {}_2F_0\left(\begin{matrix} -j, -n \\ \underline{} \end{matrix} \,\middle|\, \frac{e^{\alpha s'}}{tV(s')^2} \right)$$

We have the following relations:

$$V(s + s') = V(s) + V(s') + \alpha V(s)V(s')$$
$$H(s + s') - H(s) - H(s') = \phi(V(s)V(s')) = V(s)V(s')$$

where the first line is the addition formula for V, Lemma 2.1.1, here with $2\alpha \to \alpha$, and the second line is the cocycle identity for the Poisson system, which holds with $\phi(x) = x$, cf. Ch. 6, eq. (4.3.3).

Now we check agreement with Theorem 4.1.1. We readily identify

$$w = tV(s)V(s'), \qquad x = e^{\alpha s}\frac{V(s')}{V(s)}, \qquad y = e^{\alpha s'}\frac{V(s)}{V(s')}$$

With the relations recalled above, everything is clear via the additional identities:

$$V(s + s') = V(s) + e^{\alpha s} V(s')$$
$$= V(s') + e^{\alpha s'} V(s)$$
$$= V(s) + V(s') + \alpha V(s)V(s')$$

Here we see the interplay among the various addition formulas and special identities involved.

Now for the general Bernoulli system and the group SL(2).

4.3.2 Translations generated by X in the group $SL(2)$

For $X \in \mathrm{sl}(2)$, we have the addition formula for matrix elements of e^{sX} via Prop. 1.3.2. Using $\gamma_j = j!\,\beta^j(t)_j$ we have

$$
e^{tH(s+s')} \frac{\gamma_m \gamma_n}{m!\,n!} \left(V(s+s')\right)^{m+n} {}_2F_1 \left(\begin{array}{c} -n, -m \\ t \end{array} \middle| \frac{\pi(V(s+s'))}{\beta V(s+s')^2} \right) =
$$
$$
e^{t(H(s)+H(s'))} \frac{\gamma_m \gamma_n}{m!\,n!} V(s)^m V(s')^n \times
$$
$$
\sum_j \frac{(\beta V(s)V(s'))^j (t)_j}{j!} \, {}_2F_1 \left(\begin{array}{c} -m, -j \\ t \end{array} \middle| \frac{\pi(V(s))}{\beta V(s)^2} \right) {}_2F_1 \left(\begin{array}{c} -j, -n \\ t \end{array} \middle| \frac{\pi(V(s'))}{\beta V(s')^2} \right)
$$

Now we compare with the addition formula, Theorem 4.2.1, for the ${}_2F_1$ functions. We have

$$
w = \beta V(s)V(s'), \qquad x = \frac{\pi(V(s))}{\beta V(s)^2}, \qquad y = \frac{\pi(V(s'))}{\beta V(s')^2}
$$

Here we need the addition formula for V, Theorem 2.1.2, and the cocycle identity for H, Ch. 6, Theorem 4.2.5. We also note the identities:

$$
V(s) + V(s') + 2\alpha V(s)V(s') = V(s) - \beta V(s)^2 V(s') + \pi(V(s))V(s')
$$
$$
= V(s') - \beta V(s')^2 V(s) + \pi(V(s'))V(s)
$$

To conclude, comparing arguments of the ${}_2F_1$ functions, we find some new identities.

4.3.2.1 Proposition. *The following identities hold:*

1. For Ψ

$$
\pi(\Psi(a,b)) = \frac{\pi(a)\pi(b)}{(1 - \beta ab)^2}
$$

2. For V

$$
\pi(V(s+s')) = \frac{\pi(V(s))\pi(V(s'))}{(1 - \beta V(s)V(s'))^2}
$$

Proof: Identity #1 can be checked directly. Then #2 follows via Theorem 2.1.2, the addition formula for V. ∎

Remark. For complementary material, see Klimyk&Vilenkin[30], Vilenkin[46]. We refer the reader to Feinsilver&Schott[19] for an approach to polynomial representations and matrix elements for general Lie groups.

V. Exercises

5.1 EXERCISES

1. Give details for Proposition 3.1.2

2. Find the coefficients a_{ln}^m of Proposition 3.3.1 for the Hermite polynomials.

3. Find $\langle V^*V \rangle$, the expected value of the squared velocity.

4. Find the CSR's:

   a. $\langle V\psi_a, \psi_b \rangle$

   b. $\langle e^{tH}\psi_a, \psi_b \rangle$

   c. $\langle V^*V\psi_a, \psi_b \rangle$

5. Verify Theorem 6.3.1 of Chapter 6 using results of the present Chapter.

6. Find the coefficients c_{lm}^n of Proposition 3.3.2. for the Poisson and Gaussian systems.

7. Write out the addition formulas of §IV for $m = 0$, $m = 1$, $m=2$, for general n.

8. Fill in details of the proofs of Theorems 4.1.1 and 4.2.1.

9. Use Proposition 3.1.4 to find the composition rule

$$\langle AB\psi_a, \psi_b \rangle = (\langle A\psi_c, \psi_b \rangle, \langle B\psi_a, \psi_c \rangle)$$

   where $(\,\cdot\,, \cdot\,)$ is the inner product $(c^r, c^s) = (r!^2/\gamma_r)\,\delta_{rs}$.

10. a. Find $\langle X^2\psi_a, \psi_b \rangle$ by direct methods (e.g. as in §3.2) and by the composition rule of Problem 9.

   b. Find the 'variance' $\langle X^2 \rangle_{ab} - \langle X \rangle_{ab}^2$.

   c. Find

$$\frac{\langle X^2\psi_n, \psi_m \rangle}{\langle \psi_n, \psi_m \rangle} - \left(\frac{\langle X\psi_n, \psi_m \rangle}{\langle \psi_n, \psi_m \rangle} \right)^2$$

   d. Compare the results of parts b. and c. for $a = b$, $m = n$, giving the variance of X in the states ψ_a, ψ_n respectively.

REFERENCES

1. M. Abramowitz and I. Stegun, *Handbook of mathematical functions*, US Govt. Printing Office, 1972.

2. W.N. Bailey, *Generalized hypergeometric series*, Cambridge University Press, 1935.

3. A. Barndorff-Nielsen, *Information and exponential families*, Wiley, 1978.

4. G.G.A. Bäuerle and E.A. de Kerf, *Finite and infinite dimensional Lie algebras and applications in physics*, North-Holland, 1990.

5. J.G.F. Belinfante and B. Kolman, *A survey of Lie groups and Lie algebras with applications and computational methods*, Society for Industrial and Applied Mathematics, 1989.

6. L. Biedenharn and J. Louck, *Angular momentum in quantum mechanics and applications*, Addison-Wesley, 1977.

7. A. Bohm, *Quantum mechanics: foundations and applications*, Springer-Verlag, 1986.

8. H.P. Boas and C.R. Buck, *Polynomial expansions of analytic functions*, Springer-Verlag, 1958.

9. N. Bourbaki, *Groupes et algèbres de Lie*, Fasc. 34, Hermann, 1968.

10. L. Breiman, *Probability*, Addison-Wesley, 1968.

11. L. Comtet, *Advanced combinatorics: the art of finite and infinite expansions*, D. Reidel, 1974.

12. L. De Branges, *Hilbert spaces of entire functions*, Prentice-Hall, 1968.

13. L. De Branges and J. Rovnyak, *Square-summable power series*, Holt, Rinehart, and Winston, 1966.

14. H. Heyer, *Probability measures on locally compact groups*, Springer-Verlag, 1979.

15. Ph. Feinsilver, *Special functions, probability semigroups, and Hamiltonian flows*, Springer Lect. Notes in Math., **696**, 1978.

16. Ph. Feinsilver, Lie algebras and recurrence relations I, *Acta Applicandae Mathematicae*, **13** (1988) 291–333.

17. Ph. Feinsilver and R. Schott, Operators, stochastic processes, and Lie groups, *Springer Lect. Notes in Math.*, **1379** (1989) 75–85.

18. Ph. Feinsilver and R. Schott, An operator approach to processes on Lie groups, *Springer Lect. Notes in Math.*, **1391** (1989) 59–65.

19. Ph. Feinsilver and R. Schott, Appell systems on Lie groups, *Journal of Theoretical Probability*, **5** (1992) 251–281.

20. W. Feller, *Introduction to probability theory and its applications*, vols. I, II, Wiley, 1971.

21. W. Fulton and J. Harris, *Representation theory: a first course*, Springer-Verlag, 1991.

22. G. Gasper and M. Rahman, *Basic hypergeometric series*, Cambridge University Press, 1990.

23. R. Gilmore, *Lie groups, Lie algebras, and some of their applications*, Wiley, 1974.

24. J. Goldstein, *Semigroups of linear operators and applications*, Oxford University Press, 1985.

25. S. Helgason, *Topics in harmonic analysis on homogeneous spaces*, Birkhaüser, 1981.

26. T. Hida, *Brownian motion*, Springer-Verlag, 1980.

27. E. Hille and R.S. Phillips, *Functional analysis and semigroups*, American Mathematical Society, 1957.

28. M. Kac, Random walk and the theory of Brownian motion, *American Math. Monthly*, **54** (1947) 369–391.

29. S. Karlin and H.M. Taylor, *A first course in stochastic processes*, Academic Press, 1975.

30. A.U. Klimyk and N. Ia. Vilenkin, *Representations of Lie groups and special functions*, Kluwer Academic Publishers, 1991.

31. L.D. Landau and E.M. Lifshitz, *Quantum mechanics, nonrelativistic theory*, Pergamon Press, 1958.

32. N.N. Lebedev, *Special functions and their applications*, Dover, 1972.

33. W. Ledermann, *Introduction to group characters*, Cambridge University Press, 1987.

34. G. Letac, *Introduction to exponential families*, preprint, Université Paul Sabatier, Toulouse, 1992.

35. B.M. Levitan, *Generalized translation operators and some of their applications*, Israel program for scientific translations, Jerusalem, 1964.

36. I.G. MacDonald, *Symmetric functions and Hall polynomials*, Oxford University Press, 1979.

37. A.F. Nikiforov and V.B. Uvarov, *Special functions of mathematical physics: a unified introduction with applications*, Birkhaüser, 1988.

38. A.M. Perelomov, *Generalized coherent states and their applications*, Springer-Verlag, 1986.

39. H. Rademacher, *Topics in analytic number theory*, Springer-Verlag, 1973.

40. E.D. Rainville, *Special Functions*, Chelsea, 1971.

41. D. Revuz, *Markov chains*, North-Holland Publishing, 1975.

42. E. Ross, *Introduction to stochastic processes*, Wiley, 1977.

43. G.C. Rota (ed.), *Finite operator calculus*, Academic Press, 1975.

44. L.J. Slater, *Generalized hypergeometric functions*, Cambridge University Press, 1966.

45. G. Szëgo, *Orthogonal polynomials*, American Mathematical Society, 1975.

46. N. Ya. Vilenkin, *Special functions and the theory of group representations*, American Mathematical Society Translations, v. 22, 1968.

47. G. N. Watson, *Theory of Bessel functions*, Cambridge U. Press, 1980.

INDEX

Table of Contents

Volume 2: Special Functions and Computer Science

Table of Contents

Volume 3: Representations of Lie Groups